Data Power

"A call to arms [...] sets out a clear, persuasive argument for the need to challenge the power of platforms and systems, and details the tools to do so. A thought-provoking read."

—Professor Rob Kitchin, Maynooth University

"The first non-technical guidebook on how to live with location data and it is a truly radical response for our times. Spatial data for us, not about us."

—Jeremy W. Crampton, Professor of Urban Data Analysis, Newcastle University

Radical Geography

Series Editors:
Danny Dorling, Matthew T. Huber and Jenny Pickerill
Former editor: Kate Derickson

Also available:

Disarming Doomsday:
The Human Impact of Nuclear Weapons since Hiroshima
Becky Alexis-Martin

Unlocking Sustainable Cities:
A Manifesto for Real Change
Paul Chatterton

In Their Place:
The Imagined Geographies of Poverty
Stephen Crossley

Geographies of Digital Exclusion:
Data and Inequality
Mark Graham and Martin Dittus

Making Workers:
Radical Geographies of Education
Katharyne Mitchell

Space Invaders:
Radical Geographies of Protest
Paul Routledge

New Borders:
Migration, Hotspots and the European Superstate
Antonis Vradis, Evie Papada, Joe Painter and Anna Papoutsi

Data Power

Radical Geographies of
Control and Resistance

Jim E. Thatcher and Craig M. Dalton

PLUTO  PRESS

First published 2022 by Pluto Press
New Wing, Somerset House, Strand, London WC2R 1LA

www.plutobooks.com

Front cover designed by David Drummond for the Radical Geography series; with additional image and color work by artist, geographer, and friend Nick Lally. Image source: "2020.06.06 Protesting the Murder of George Floyd, Washington, DC USA 158 20209" by tedeytan and is licensed with CC BY-SA 2.0. The color scheme, viridis, was created by Stéfan van der Walt and Nathaniel Smith and is intended to more clearly visualize data for readers with common forms of colorblindness.

British Library Cataloguing in Publication Data
A catalogue record for this book is available from the British Library

ISBN   978 0 7453 4007 4   Paperback
ISBN   978 0 7453 4008 1   Hardback
ISBN   978 1786805 56 0   PDF
ISBN   978 1 786805 57 7   EPUB

Typeset by Stanford DTP Services, Northampton, England

Contents

List of Figures and Tables   vi
Series Preface   viii
Acknowledgments   ix
List of Abbreviations   xi

Introduction: Technology and the Axes of Hope and Fear   1

1   Life in the Age of Big Data   13

2   What Are Our Data, and What Are They Worth?   46

3   Existing Everyday Resistances   65

4   Contesting the Data Spectacle   84

5   Our Data Are Us, So Make Them Ours   119

Epilogue   131

Notes   133
Bibliography   140
Index   159

Figures and Tables

FIGURES

1.1 The Lackawana Valley, by George Inness   14
1.2 An 1863 image of Philip Reis' telephone from the
German newspaper *Die Gartenlaube*   20
1.3 Images at two scales around Boston from the "One Dot
Per Person for the Entire United States" visualization
created by the Demographics Research Group at the
University of Virginia   35
1.4 A clearly demarcated US military base discovered in
Strava's Global Heatmap by Nathan Ruser   37
1.5 Inspired by Nikita Barsukov's work, Nathan Yao built
these maps using public RunKeeper data   44
1.6 Sample code to scrape RunKeeper's public routes for
the city of Tacoma   44
2.1 One of Chicago's police surveillance cameras   60
3.1 A map of surveillance cameras in Times Square,
Manhattan, created by the Surveillance Camera Players
in May 2005   80
4.1 Le Corbusier's Plan Voisin model for the redevelopment
of Paris   96
4.2 Guy Debord's *Life Continues to Be Free and Easy* (1959)   97
4.3 Map showing one of Precarias a la Deriva's drifts through
the daily lives and practices of domestic workers in Madrid   100
4.4 Walking route, Thursday, September 1, 2016   103
4.5 "Cat and Girl are Situationists," by Dorothy Gambrell   106
4.6 A billboard in San Francisco détourned by the Billboard
Liberation Front, a group devoted to "improving outdoor
advertising since 1977"   107
4.7 A détournement of corporate statements made in support
of Black Lives Matter, created by Chris Franklin   109
4.8 A screenshot of Inside Airbnb's web map, created by
Murray Cox and Inside Airbnb   111

4.9   A screenshot of the Anti Eviction Mapping Project's
      Mapping Relocation map   112

TABLES

3.1   A typology of responses to data capitalism   71
5.1   Suggested yarn colors for a temperature blanket based on
      Tacoma, WA   129

Series Preface

The Radical Geography series consists of accessible books which use geographical perspectives to understand issues of social and political concern. These short books include critiques of existing government policies and alternatives to staid ways of thinking about our societies. They feature stories of radical social and political activism, guides to achieving change, and arguments about why we need to think differently on many contemporary issues if we are to live better together on this planet.

A geographical perspective involves seeing the connections within and between places, as well as considering the role of space and scale to develop a new and better understanding of current problems. Written largely by academic geographers, books in the series deliberately target issues of political, environmental, and social concern. The series showcases clear explications of geographical approaches to social problems, and it has a particular interest in action currently being undertaken to achieve positive change that is radical, achievable, real, and relevant.

The target audience ranges from undergraduates to experienced scholars, as well as from activists to conventional policy-makers, but these books are also for people interested in the world who do not already have a radical outlook and who want to be engaged and informed by a short, well written and thought-provoking book.

Danny Dorling, Matthew T. Huber, and Jenny Pickerill

Series Editors

Acknowledgments

Change requires hard work and building solidarity. With the help of so many others, we've written this book as an exercise on building alternatives.

Books are weird, complicated things in which there are fragments of both ephemeral conversations and decadal arguments. For Jim, this book would have been impossible without the support of family and friends near and far. I'd like to first thank Courtney, my partner for nearly two decades now, for her support, calm, and general equanimity. Hannah, age eight, and Ben, age four, were similarly crucial for the development of this book, as were my parents, Sally and Richard, for their periodic care of those two. Intellectually, I owe a debt of gratitude to Craig for our many collaborations, but also to folks including Luke Bergmann (who provided notes on an early draft), Nick Lally (who revised the cover), Laura Imaoka, Clancy Wilmott, Emma Fraser, David O'Sullivan, Dillon Mahmoudi, Alicia Cowart, Ryan Burns, Kelly Kay, Alida Cantor, Chris Knudson, Carolyn Fish, Anthony Robinson, Danny Kunches, David Retchless, Lauren Anderson, Megan Finn, Madelynn von Baeyer, Josh Gray, Karen Thatcher, and—I am certain—a host of individuals I'm unintentionally forgetting.

Also, I appreciate the many master's students with whom I've had long-standing discussions on these topics, including Corrine Armistead, Matt Seto, and Ryan Mitchell. Although gone now, David Waring deserves mention for our many late-night conversations over nearly 20 years of friendship. A special, long-term, thanks must go to my erstwhile advisor, James McCarthy—any misapplication of rigorous Marxist thinking should not reflect poorly upon the source. Finally, I must thank Joe, Jon, Dan, Steve, Jason, and Mike for Friday evening gaming; we'll make diamond in Rocket League someday.

This book was also only possible with the support of many people in Craig's life. This includes both those who directly inform the ideas of my work and those whose reproductive labor and support made it possible to develop and write a book at all while simultaneously raising a child and teaching heavy loads every semester though a pandemic. Cecilia, I cannot imagine this book without your unwavering emotional support, material labor, down-to-earth feedback, and patience under exceptional circumstances. Thank you. Rowan, for consistently bringing me back to what really matters in life and the true magic of the everyday. Thank you as well to Courtney for supporting Jim and keeping him going; I see and appreciate your work from 2,800 miles away.

Gma, Papa, Momo, Teetee, GrAnna, Grandpa, thank you for all the love and childcare while we podded up in one house or another and the COVID positivity rates outside kept rising. Dad, decades ago I asked you why you wrote a book if you would not be paid much for it. You said, "To change how people think." We'll see how this one goes.

To all the members of the Counter Cartographies Collective, you taught me the true meaning of resistance as well as counter-mapping, the Situationists, and to pay attention to reproductive labor. Y'all continue to inspire my ideas. In particular, thank you Maribel Casas Cortés for introducing me to the work of Precarias a la Deriva.

This book would not be half as strong without the direct input of Tim Stallmann and Debra Mackinnon early in the process. Thank you as well to Clancy Wilmott, Emma Fraser, Ryan Burns, Leah Mesiterlin, Gregory Donovan, Liz Mason-Deese, and Erin McElroy for our productive conversations around these themes. Kari Jensen, Grant Saff, Zilkia Janer, your support provided a foundation at Hofstra that made this scholarship possible. And above all, thank you Jim, for getting me to dream bigger.

Finally, from both of us, this book was only possible through the incredible patience and attention to detail of our exceptional editor, David Castle, the hard work of everyone at Pluto, as well as the anonymous reviewer.

Abbreviations

3Cs	Counter-Cartographies Collective
3PLA	Third Department of the People's Liberation Army's General Staff Department (China)
AEMP	Anti-Eviction Mapping Project
AI	artificial intelligence
API	Application Programming Interface
CCPA	California Consumer Privacy Act
CCTV	closed-circuit television
CEO	Chief Executive Officer
EU	European Union
EULA	end-user license agreement
FCC	Federal Communications Commission (US)
GDPR	General Data Protection Regulation (EU)
GML	Generalized Markup Language
GPS	Global Positioning System
HFT	high-frequency trading
IoT	Internet of Things
JSON	JavaScript Object Notation
NSA	National Security Agency (US)
SGML	Standardized Generalized Markup Language
VPN	virtual private network
XML	Extensible Markup Language

Introduction
Technology and the Axes of Hope and Fear

Our first step is to bring back curiosity.

(Tsing 2015, 4)

We are the most fantastic and beautiful mistake.

(Russell 2020, 147)

One step after another, each recorded and located by the Global Positioning System (GPS) and shared with the world. Sequential steps repeated daily in our morning run or commute become part of an economic cycle of digital tracking, extracting our location data and serving parts back to us as directions, as ads, as insurance rates. And also as egregious privacy violations which set off, like clockwork, another cycle—a media cycle. In 2018, Nathan Ruser revealed that Strava's Global Heatmap of users' exercise routes had inadvertently revealed the locations of several nominally secret military bases. A parade of news articles followed that ranged from how-to pieces on managing the fitness application's privacy settings (Pardes 2018) to more widely questioning the very concept of privacy and informed consent (Tufekci 2018).

The problem with this media cycle is not with any individual piece of content. Pardes' *WIRED* article is an excellent guide to navigating Strava's privacy and security settings. Rather, the problem lies in how each data scandal is framed as separate and surprising, seemingly unforeseeable even as each extraction of data for the purpose of profit inevitably sets up the conditions for exactly this kind of event (Thatcher 2018). Even a cursory glance at recent technology news reveals the cyclical nature of such spatial (geographic/location) data abuse narratives: before Strava there was Microsoft's Avoiding the Ghetto patent (Thatcher 2014), and before that, Girls Around Me leveraged the Foursquare and Facebook data to help men stalk women (Bilton 2012). Examples of outrage, and even congressional and European Union (EU) court hearings (Jacobson 2020), abound, but policy is slow and at times reversed or co-opted by the companies it is meant to regulate. As with running for exercise, the destination of this data cycle isn't the point; maintaining the cycle is. Continuing to extract our data themselves,[1] and spatial data in particular, is profitable. While an individual application or feature might change due to a data scandal, the overarching cycle of spatial data creation, extraction, and exchange with little regard for the users producing the data or other consequences continues apace.

This is a book about what we can do to change that.

Non-fiction narratives about technology tend to be either utopian or dystopian: eschatological visions of mobile applications ending pandemics or of drone strikes silencing political dissent. Accounts of Google's attempted smart-city neighborhood in Toronto or Cambridge Analytica make for great stories, but they miss the forest for the trees. Both tropes oversimplify complex processes and contexts, hamstringing attempts to understand how individual cases reflect broader systems. Processes of profit-seeking and capital accumulation frame recent discussions around technology, delimiting what is thought possible and desirable for technology to do. *That need not be the case. More alternatives are possible.* We explore hopeful tactics and strategies for living amidst and moving beyond the ruins created by an ideology of technology which "move[s] fast and break[s] things" (Facebook founder and Chief Executive Officer Mark Zuckerberg, quoted in Taneja 2019). In recent decades, technology firms sought to "disrupt" existing social relations and remake them in their own image: Facebook with friendship, Uber with movement, Google with knowledge. In so doing, ever greater parts of daily life, of personal identity, become the playground of this speculative form of capitalism.

Writing in the *Harvard Business Review*, Hemant Taneja (2019) declares that the era of new tech entrepreneurship is over, due to people growing weary of the cavalcade of abuses perpetrated by firms like Facebook and Google. Taneja and others rightly note that this is in part due to a new generation of technologies, such as GPS tracking, facial recognition, and genomic profiling, that are far more personal and tangible in the daily lives of individual people. And, as such, a number of institutes, initiatives, and other public-facing endeavors have emerged to engage with the underlying algorithms and biases of these systems. These initiatives operate at different scales and with different stakeholders. Some, such as the Partnership On AI, are explicitly aimed at working with the leaders of tech firms to shape practices. Others, such as the Electronic Frontier Foundation, employ lobbying and lawsuits in attempts to shape policy. There is nothing inherently wrong, and many things truly beneficial, about such approaches. Nevertheless, we choose not to focus on institutional policies, real or potential, in this book.

We focus on lived cultures of technology, especially the grounded experiences and potential geographies of resistance amidst the everyday. We focus on a praxis—the putting into action, the embodying, of the hard-won insights of theory—to live joyfully within the social and literal ruins of data capitalism. Anna Lowenhaupt Tsing (2015) brilliantly depicts the communities that grow up amidst environmental change and degradation (the ruins of another aspect of capitalism) and how they fit within larger global systems of production and consumption. We seek to do the same amidst the creation, extraction, and analysis of the spatial data produced through everyday life. We focus upon a daily life enmeshed within the technical apparatuses of new data regimes as a means of being "truly present … as mortal critters entwined in myriad unfinished configurations of places, times, matters, meanings" (Haraway 2016, 1). Our language and examples are attempts to describe, as new media theorist McKenzie Wark (2020, 168) puts it, "a present that could be open to other futures." Through this, we offer a guide on what is to be done to live with, not under, new spatial technologies.

In this book, we offer five chapters centered around the goal of finding new and creative ways to enact a radical political praxis with new spatial technologies. Each chapter can be read independently, but together they form a hopeful narrative arcing towards something better. The remainder of this chapter introduces the book in detail: first, answering for whom this book is intended; second, outlining the content of each subsequent chapter; third, reflexively examining the perspective of the authors' and what is therefore missing from this volume; and finally, providing a brief note on how to make use of the online resources associated with the text.

SO WHAT? WHY THIS BOOK MATTERS

At this point, you've picked up and opened a book from Pluto Press' Radical Geography series titled *Data Power*. From our perspective, it seems safe to assume that you have at least a passing interest in the role of data and technologies in culture and society, radical geography's ideas and practices, or the critical orientation of Pluto's publications. If that is the case, then this book is for you.

This book does not assume any deep familiarity with technical systems or digital spatial data's creation, dispossession, and commodification as we introduce what is important and relevant over the course of the chapters. That said, it is useful to begin here with a brief description of the production of data through everyday living as a key means of capitalist profit-seeking. The book centers on these social processes and how they may be resisted.

Put simply, each interaction with a digital system produces data. For example, a post to Facebook will often record not simply the post itself, but also where the user was when it was made, other users who were with them, and even the browser/device used to make the post (this is often referred to as the post's metadata). These data are extracted, tied to other data points about the user, and analyzed to produce digital representations of them. These representations are not complete captures of their life, but specifically focused on predicting their actions and, in particular, their consumptive practices. This is why Craig receives advertisements for a new tent after mentioning camping on Twitter.

While each individual data point holds little meaning or value, a collection of data points tends to mean more than the same points in isolation. Data are much more valuable when connected to still more data. The degree to which multiple data points can be tied together, especially if it can anticipate how likely (or able) a user is to spend money, determines the value of the data they produce for advertising and data analytic firms like Facebook, Google, Twitter, and many others. The massive scales of data and users at which such data technology companies operate produce centralized, multi-billion-dollar industries that continually seek to shape your actions, your life, in ways more amenable to predictable consumption, and therein, their bottom line.

In light of these ongoing processes, this book does three things. First, it surveys the current context in which new technological regimes, historically contingent socio-political systems, of spatial data play in our world. Rather than the one-off solutionism offered by technology's most disruptive boosters, we instead "move slow," purposefully turning towards the long history of critical and radical thought concerning the questions technology poses for our lives, from Walter Benjamin to Ruha Benjamin. We find these ideas to be the best available tools for "staying with the trouble" (Haraway 2016), situating what we know about the present and sifting through the past to help create alternative futures. Second, we examine current individualized responses to data regimes, demonstrating both where these practices succeed and where their limitations lie. Third, we outline a clear set of collective practices focused on developing radical solidarities, kinships against and through spatial technologies. Our purpose is to develop ways of (re)asserting our humanity within the sociotechnical milieu in which we live—a guidebook of sorts for living *with* data. Perhaps these pages will even be as "disruptive" as the unicorns and rock stars of the tech industry dream themselves to be.

What we promise is that by the end of this book, you will have a deeper perspective on why and where the location data you create in your day-to-day life are extracted, analyzed, and come to stand for you as well as a conceptual toolkit for evaluating, resisting, and making use of those systems more on your own terms.

THE REST OF THIS BOOK: A DIALECTIC TENSION

In the past, we've written about the ongoing framing of technology as a double-edged sword with respect to culture and society (Thatcher et al. 2018, xvi). Following that, we structure the rest of this book as a "broad dialectic between hope for technology's role as liberator and fear of its domination of everyday life" by drawing on Kingsbury and Jones' (2009) interpretation of the Frankfurt School's studies of technology. We return to that somewhat obscure, critical social theory not as a means of obfuscating our point, but of developing our practice.[2] Whether or not you're familiar with the Frankfurt School, the tension of technological hope and fear is a common one in current culture. One example is the question of labor: "Robots will destroy our jobs—and we're not ready for it" (Shewan 2017) versus "Robots are increasing our wages, not stealing our jobs" (Clark 2015). Thinking through these ideas dialectically allows us to understand how *both* can be and likely are true, but also that both are beside the point unless we radically alter the ways technological and social relations interact to reassert our humanity.

Thinking through those ideas appears here via the themes explored in each chapter. In Chapters 1 and 2, we follow technology and society from the nineteenth century up through our present day; in doing so, we tie the ruins in which we live not to Silicon Valley's promised solutionism, but to the long-standing roles quantification, classification, and abstraction have played in capitalist exploitation. Chapter 3 then surveys current forms of individual resistance and acceptance, of technological enabling and constraining actions, that have occurred with and in response to spatial data regimes. Finally, Chapters 4 and 5 build from the limits found in current practices to develop collective modes of resistance and a synthetic set of concrete practices that engage new systems of daily data creation and collection in ways that produce new solidarities and hopeful experiences. Finally, in a brief epilogue, we reflect upon the limits of our own scholarship and how the writing of this book has been shaped by an ever-changing world.

We begin the first chapter, "Life in the Age of Big Data," by stepping backwards in time to the widespread adoption of telephones and motion pictures. By puncturing the liberal myth of individual empowerment through technology, we draw on an argument that begins with the work of Karl Marx and works towards the present moment in which data have come to represent and speak for us. The works of Walter Benjamin and Guy Debord figure prominently in this chapter, as do other critical theorists that allow us to draw parallels between early critiques of cultures of technology and current experiences of mobile phones and media today. Specific parallels are drawn between how Benjamin understood the telephone as reaching into and disrupting personal private spaces and how the spatial data produced by mobile devices now inscribe and map our most intimate moments.

Refusing to remain dormant in a misplaced past, we follow this line of thinking (and critiques thereof) up to the present moment, diagnosing the ways in which spatial data and their analyses influence and sustain social relations and produce new spaces of consumptive experience. This view of how new spatial data systems affect where we go, who we encounter, and what we can (and cannot) know of places and individuals stakes out the fear side of the hope/fear tension.

Throughout, current cases such as Girls Around Me and Strava complicate and ground our arguments in a critique of the everyday. The inequities of these experiences and the resulting media coverage, as well as the assumptions of both the creators and users of these systems, reveal the limits of popular critique and the perennial framing of new privacy scandals as unexpected. They also lay clear the multiple social and physical scales at which these systems operate with and upon our lives.

Expanding from the individual scale, the second chapter, "What Are Our Data, and What Are They Worth?", moves towards societal narratives around data and algorithms. We introduce the concept of data colonialism and the "wild west" roots of the Silicon Valley ethos, calling particular attention to the ways in which we are dispossessed from the data we produce. From there, we demonstrate the near-theological, faith-based relationships modern society has with data: the belief that data, their analyses, and their visualization will progressively, inevitably, and irrevocably improve our lives. Second, we argue that through these individual data experiences and societal narratives, our personal relationships with data have become bearing witness to, rather than actively intervening in, the construction and analysis of the data that represent us.

Early in the chapter, we draw on work that demonstrates how faith in the data, analyses, networks, and infrastructures that support algorithmic decision-making have become the dominant metaphors for society in the twenty-first century—*data is the new oil, we must run our cities like start-ups*, and so on. Returning to examples from the popular press, we focus first on Silicon Valley and the narratives that have emerged around its most prominent corporations. We also highlight the portrayal of similar cultures of technology in China as a constructed dystopian other. The faith-based nature of these narratives articulates a desire for an ever smoother, more predictable form of societal organization. Such narratives contrast with the actual lived experiences of the individuals who create and then are separated from and represented by data under these regimes. Building on the work of Melissa Gregg (2015), we untangle this process of data spectacle.

The data spectacle in Chapter 2 marks a turning point in the book as it begins to lay the groundwork for material practices to address the problems with current data regimes. Focusing on spatial data, their creation, extraction, and analysis across scales as key aspects of our daily lives, we now turn towards active engagements with and resistances to data relations and geographies.

The third chapter, "Existing Everyday Resistances," evaluates a series of existing tactics for individuals resisting the dispossession and commodification of data. We introduce a typology for understanding how individuals engage daily data production and analysis through acts of *acceptance, resistance, making present*, and *escape*.

By *acceptance*, we mean consenting or acquiescing to a technology's terms of service to function and participate in society and culture today. The pretense of these terms allows data regimes to slip into invisible ubiquity, disappearing into the banal plain sight of everyday life. Under these circumstances, they re-emerge for conscious consideration and critique only at moments of rupture where systems break or data leak.

Resistance therefore constitutes tactics that contest the data production and extraction outlined in those terms. For example, active resistance may create inaccurate or false tracking data and insert them into targeted marketing systems. *Making present* is another resistance tactic that reveals the invisible infrastructures underlying regimes of data accumulation, such as the mapping of data centers or reverse engineering and making public various sorting algorithms. Finally, *escape* refers to attempts to remove oneself from the generation of spatial data at various levels of intensity, whether living entirely off-grid or simply switching to an ancient flip phone (or none at all). Opportunities for all of these tactics, but escape in particular, are not equitably distributed throughout society.

The purpose of this typology is to provide a shorthand means of assessing the intent, effectiveness, and limitations of existing practices. The focus on examples from artists and academics is not by coincidence, but highlights the avant-garde and individualistic nature of many of these practices. Academics are, if nothing else, excellent neoliberal subjects—just look at our citation rates. Throughout Chapter 3 we tie together common threads between these approaches and the situations in which they arose. We call for a weaving together of disparate attempts into a lived practice of the everyday that makes new technological systems fundamentally work towards the building of solidarities and liberation of humanity.

In Chapter 4, "Contesting the Data Spectacle," we answer this call with a focus on collective modes of resistance. Returning once more to the asymmetries between the individual data producer and the firms which extract and analyze those data, we again confront the data spectacle. We offer four approaches informed by social theory by which groups or coalitions may confront and change the data spectacle: data regulation, data dérive, data détournement, and data strikes.

With the European Union's General Data Protection Regulation (GDPR) and similar changes coming online elsewhere, regulation of individuals' spatial data looks promising thus far. However, even the GDPR is unlikely to overturn the capital relations behind data-driven firms, and even the strongest policies have limited social effects without parallel cultural changes. With those limitations in mind, we explore the *data dérive* (drift). This builds on the dérives developed by the Situationists, an international radical art and political collective most active during the 1950s–1960s in France, and Precarias a la Deriva, a radical feminist collective for women active in Spain in the 2000s. Building on their work, we propose applying the dérive to the current contexts of the data spectacle. The resulting "data dérive" is a way to purposefully approach spaces of everyday life while cognizant of associated data regimes (Thatcher and Dalton 2017). Building on the dérive, we next engage another Situationist approach to develop what we call *data détournement*. This approach involves employing or differently applying data not as a commodity, but as a means for political change. Counter-mapping provides exciting examples of this sort of practice. For the last mode of collective resistance, we briefly propose a *data strike*. The idea is to withhold data, to the extent possible, from data-driven companies to incentivize them to make changes. We say strike, not boycott, because producing data, and therein value, is labor.

Chapter 5, "Our Data Are Us, So Make Them Ours," returns to the idea that the data regimes in which we find ourselves are not wholly "new" and have their roots in long-standing processes of exploitation and domination within modern, capitalist societies. For example, the direct development of modern locational tracking for targeted advertising springs from earlier geodemographic profiling and anti-poverty initiatives (Dalton and Thatcher 2015; Eubanks 2018). History matters, as examples of taking back or repurposing data indicate some of the radical moves available to us today.

Eschewing blanket rejections of the role of technology in our lives, we return to the question of liberation, of asserting our humanity *with* and *through* technology. Based on the critical and empirical examples throughout the book, we argue that we can both anticipate and understand the role data and their analyses play in enabling and constraining our everyday spaces and knowledges. In so doing, we are able to contest, repurpose, and recreate these spaces through a set of concrete practices that inform a radical politics of change. We end the book with three "calls to action" that form the basis of a more engaged, technologically informed radical politics.

The first is a rejection of the "Who could have known?" fictions that perpetuate modern popular press coverage of the media. The second is a return to the examples from Chapters 3 and 4 to once more suggest where and how we might find working solidarities in a world of data-derived value. Third, we suggest a praxis that "lives in the cracks," one that embraces incomplete knowledge, partiality, and the subversion and repurposing of technology in novel ways. We provide examples of this through a living repository where readers are encouraged to contribute and comment—a perpetual work in progress. Without denying the oppressive nature of technology in society, we conclude with an invitation to act: *we can and must reassert our shared humanity, not just in the face of new regimes of data creation, extraction, and analysis, but through the very technologies which make these regimes possible.* We must create new spaces of affinity, new politics of change *with and through* the enabling elements of new spatial technologies while eschewing, resisting, and subverting the constraining ones.

Finally, in a brief epilogue, we reflect on the production of this book itself—how it shifted and changed, both due to our own scholarship (and limits thereof) and the world around us. COVID-19, the presence of which is felt throughout the book, also shaped its construction—who could (not) be interviewed, when, and how. Although different in form and intent, we draw inspiration for this reflection from the audit Catherine D'Ignazio and Lauren F. Klein (2020) conducted at the end of their own excellent work, *Data Feminism*.

WHO SPEAKS FOR WHOM?

The contours of data accumulation are uneven in both existing and potential spaces. What we can and do know of a young stockbroker in London who uses her mobile phone to make trades, find restaurants, and summon ride-sharing services is different from what we may know of a farmer in sub-Saharan Africa who uses a shared mobile phone to coordinate market prices and check upcoming weather forecasts. Furthermore, the dangers and risks at which our own data might put us are highly variable upon our position and privilege.

The UK Home Office's use of data collected by a charity on "hard sleepers" (homeless people) illustrates this point.[3] Data intended to ensure that case workers spoke each individual's native language was instead repurposed as a means of locating, apprehending, and deporting immigrant homeless individuals. For the same reason, while mapping the daily practices of migrant workers in North Carolina could provide deep insights into their needs, it would also create a detailed record for deportation purposes (Dalton et al. 2016). The techniques we discuss, from active resistance and making present to data dérives and data strikes, take many forms and will depend on context. Thus, their potential risks and rewards are highly situated and variable. Our intent is never to suggest that any reader must follow any given recommendation or that to not do so is some kind of moral failing. Depending on the circumstances, some modes of resistance may not be appropriate because under those circumstances they are dangerous or likely to harm other people, or they simply may not be possible.

The practices and ideas we develop ultimately stem from our own positionality. We are cis white men raised and educated in the global north. We each hold professor positions within neoliberal universities and live in urban areas. While our experiences differ, they are also limited. While we bring in examples of spatial data regimes and resistances from around the globe and ideas from different kinds of thinkers, there are blind spots and aporias in our work, just as there are in any scholarship. We say all this to make clear from the outset that as reflexive as we try to be throughout this work, we recognize those efforts are never complete. Indeed, working with individualized data offers new opportunities to connect and entwine subjects and objects, such as their own spatial data, for better-situated thinking.

Our goal is to build shared affinities, tactics, and solidarities through experiences of new spatial data regimes that are sorting and oppressing, enabling, and constraining our actions. To do so, we draw upon our decades of experience researching and teaching geospatial technologies at the undergraduate and graduate levels; but rather than eschew our limitations, we call attention to them now and ask that readers keep their own perspectives in mind as they read.

ONLINE RESOURCES (OR THE DIGITAL APPENDIX)

New technologies are rarely as "disruptive" as they seem and never as much as their boosters claim. Throughout this book we emphasize a need to view technology as part and parcel of larger historical systems of capitalist exploitation and development. Our path forward is shaped by the successes and failures of past resistances and past solidarities. Nevertheless, how a specific technology functions, its mechanical and algorithmic processes, can and will differ over time. For example, around the turn of the century, Extensible Markup Language (XML) was the "hot" new way to structure data for sharing and use across the internet.[4] Writing 20 years later, XML is viewed as relatively archaic. Software developers today favor formats like JavaScript Object Notation (JSON), which forms the backbone of many spatial data applications in its geoJSON form.

Putting aside the alphabet soup of acronyms, we can't answer now what specific apps, data, and algorithms you will encounter in the years after this book is published. Acts that work to subvert and resist the current Snapchat application may be meaningless against whatever TikTok releases next (and in years to come, those application names will likely be meaningless or antiquated to our readers). In order to provide a meaningful praxis, one informed by long histories and attuned to the present moment, this book has a digital companion.

At https://github.com/DataResistance you will find a continually updated set of digital resources that extend, update, and revise the content of this book. As a living archive, you will find, for example, code for pulling and visualizing your tracking data from Google Maps as well as links to ongoing projects by other artists, academics, and collaborators. That site serves and will continue to serve as a generalized repository for tools and ideas on how to speak with our digital data. It is intended as a collaborative space for discussion as well as contribution, and all of our readers are encouraged to visit, collaborate, and perpetuate a resistive reclaiming of technology.

In the chapters that follow, we chart the history, present, and future of spatial data and the devices which create, extract, and analyze it in our lives. Together, they develop practices that we can carry through our everyday lives, describing not only what is, but more importantly, what might and must be. Spatial data stand for us, but we must learn how to make them our own.

1

Life in the Age of Big Data[1]

Not many of those who use the apparatus know what devastation it once wreaked in family circles. The sound with which it rang between two and four in the afternoon, when a schoolfriend wished to speak to me, was an alarm signal that menaced not only my parents' midday nap but the historical era that underwrote and enveloped this siesta.

(Benjamin 2008[1938], 77)

[Chaplin's] unique significance lies in the fact that, in his work, the human being is integrated into the film image by way of his gestures— that is, his bodily and mental posture. The innovation of Chaplin's gestures is that he dissects the expressive movements of human beings into a series of minute innervations. Each single movement he makes is composed of a succession of staccato bits of movement. Whether it is his walk, the way he handles his cane, or the way he raises his hat— always the same jerky sequence of tiny movements applies the law of cinematic image sequence to human motorial functions.

(Benjamin 2008[1935b], 340)

It matters which stories tell stories, which concepts think concepts.

(Haraway 2016, 101)

How is technology impacting society? How can we live amidst technology? Far too often, the term "technology" serves as a catch-all, an inevitable, ineffable force behind change. Even when closely evaluated, technology's inner workings and effects are still often presented as an unexplained or proprietary black box—if not inevitable, a fait accompli justified by self-interested claims to innovation and newness. But these are not new questions or new claims, and in reality, a technology's impacts reflect the social imperatives of its designers and users, not a force of nature.

Picking apart the promises of technology to find alternative paths requires an understanding of the history of their functions and

Figure 1.1   The Lackawana Valley, by George Inness. (National Gallery of Art, public domain)

myth-makings within capitalism. The questions of how data are used today and how to resist or change associated relationships require engaging ideas about the roles technologies play in societies and the opportunities opened therein. In the case of data, this involves ongoing processes of quantification and representation that undergird technological data produced by and for systems of capital accumulation.

Due to this connection to capitalism, theorists of technology usually emphasize the first question, *how data are used*, and examine the consequences of technologies for societies at large and the subsequent roles of people within them. Their work provides a critical foundation on which to unpack what technology actually is and how it actually works, from Marx's hand-mill to Adorno's cinema to Marcuse's one-dimensional man to Feenberg's instrumentalizations. But unlike those theorists, we are more concerned with the second question, *how the current relations of data might be resisted or repurposed*, a careful exploration of possibilities for how we can live with data-driven technology in more practical, equitable, but no less critical, ways. We necessarily focus more on particular, situated experiences of the everyday rather than more broad attempts to theorize technology and society as a whole.

One of the first to evaluate such technologies in conjunction with changes in daily life was culture critic and theorist Walter Benjamin. Active from the 1920s to the 1940s, his ideas went on to shape the post-war Frankfurt School of Critical Theory and numerous scholars, artists, and activists thereafter. In the passage on the telephone, he calls attention to the intimate disruptions of technologies that are now banal, tying them outwards to profound shifts in societal relations ("an alarm signal that menaced ... the historical era"). Further, with Chaplin he illustrates how people can adapt themselves to technologies designed by others ("applies the law of cinematic image sequence to human motorial functions"). In each case, human and technology couple in intimate relations involving both fear and hope.

In this chapter, we lay the conceptual foundation for how current relations of data may be resisted and better lived using critical ideas about technology, culture, and capitalism. We begin by dispelling the liberal myth that technology (and data) is always neutral and simply reflects the priorities of those who use it. We also confront the idea that technologies determine economic relations and culture. The historically situated reciprocal relationships of societies and technologies are far more complex than these linear formulations. Shifting from technology in general to representational, for example data-driven, technologies, we trace how pre-war Walter Benjamin saw hope for class consciousness and liberation where his post-war colleagues, Horkheimer and Adorno, found none. But given the failure of Benjamin's hopes, how can we resist the current cultural economy and the data technologies that facilitate it? We explore three proposed ways out through everyday practice from Heidegger, Debord, and Marcuse. Finally, building on all these ideas, we evaluate two recent geographic data scandals, Strava and the Home Office's use of homelessness data.

THE (NEO)LIBERAL MYTH OF TECHNOLOGICAL EMPOWERMENT

When technology isn't referred to as an external, natural force, one common trope assigns all of its consequences to the choices of individual users. Such claims rest upon the idea of technology as inherently neutral, its effects instead decided by the person who wields it. Obviously, this is true to an extent; a sword can be used for cutting, threatening, or repurposed into a plowshare; so too may an encrypted email contain directions for a romantic getaway or the plans for an improvised explosive device.

However, this formulation entirely misses the social and material contexts of those actions that lead to and delimit the options of those making the choice. *We make use of our technologies, but not in circumstances of our own choosing.*

First, people cannot act outside their social, cultural contexts. Focusing on someone's choice to use a sword ignores the systems of power that put them into that position. A conscripted footsoldier directed to use that sword based on commands from a king under pain of execution for disobeying orders doesn't have many choices. Moreover, social contexts go far deeper. The very basis of social systems and ways of knowing technologies are socially constructed, be they the militaries of medieval kingdoms or the social pressures placed upon young people to use social media "appropriately": "Delete those pictures of you partying and make some LinkedIn connections or you'll never land that dream job!"

Second, the designed, material structure of a technology has consequences for how someone may use it. Designers create technologies to fulfill particular social imperatives, such as capital accumulation, national defense, or fun. As a result, the literal construction of those technologies reflects the purposes and biases of their designers. One of the reasons so many early mobile applications focused upon the interests of men with high levels of disposable income in urban settings is because they were designed by white men with high levels of disposable income living in an urban environment and who therefore knew what to create for that demographic group. UnTappd, a social network built around tracking craft beer consumption, is emblematic of this.

However, a technology's designer does not wholly determine subsequent users' actions. A user may apply a technology in a way not foreseen or intended by a designer, but that user's range of possible actions, "margin of maneuver," is delimited by the technology's material structure, and therein, the designer's social imperatives (Feenberg 1999). The designers' social imperatives:

> create a framework of activity, a field of play, but they do not determine every move …. The "weaker players," those whose lives or work are structured by the technical mediations selected by management, are constantly solicited to operate in this range of unpredictable effects.
>
> (Feenberg 2002, 86–87)

A small screwdriver may be used on screws, or as a lockpick, but it is not useful as a hammer, no matter how hard you swing it.

This is also the case in purely digital technologies. Google Maps is a consumer-grade navigation service, and excels in that role. It's also been applied to purposes within its material margin of maneuver that its original designers did not anticipate, from real estate to contemporary art (Dalton 2015). However, it's a poor tool beyond that margin, such as for commercial truckers concerned about the heights of freeway overpasses, regional planners who want to evaluate the environmental impact of new zoning rules, or land trusts trying to create more affordable housing.

Moreover, due to social and material context, the intentional and unintentional effects of technologies are not equally distributed. Technologies frequently reinforce and reproduce social biases, often in new, powerful ways. Ruha Benjamin (2019b) writes at length about what she terms "Jim Code": how digital technologies facilitate and reinforce racist ideas and practices, not only in intentionally white supremacist social media, but also in systems intended to be impartial. She harrowingly depicts how the data fed into algorithms meant to establish recidivism rates for convicts feed upon the unequal distribution of racial justice within existing structures. It is not simply Garbage In, Garbage Out, but Racism In, Racism Out. As geographer Brian Jefferson (2020, 6) notes, "computation does not signify a new cultural logic so much as it performs an upgrade of entrenched modes of social differentiation and dominance." The designers of such technologies may not intend their works to have such racialized consequences, but by performing technological design within larger systems of racialized capitalism, such unequal consequences are both extremely common in practice and perhaps unavoidable in execution. Political geographer Louise Amoore (2020a, 146), expands on this idea:

> This means that one could be doubtful of all claims, for example, that the bias or the violence could be excised from the algorithm and begin instead from the intractable political relations between the algorithm and the data from which it learns.

The biases of a technology cannot be fixed unless the circumstances from which it springs and within which it functions change.

In sum, technologies are part of situated, recursive processes between social and material aspects which produce one another. Technologies shape our options and actions as we, in turn, shape both the structure of technologies and what we are expected to do with them. These relations are inherently part of technologies and their use. A simple recidivism algorithm written in San Francisco is intimately tied to the history of racism within the United States judicial system. Uber's allocation of rides connects the gig economy to the history of labor exploitation. To follow Haraway (2016), staying with the trouble means we must situate our approach. To look at where we are, and where we might go, we must first turn to how we got here. This is not in terms of abstract data and technology as entities separate from people, but as data and technology produced and used and repurposed by and for people amidst an actual history and place.

POLITICAL ECONOMY AND TECHNOLOGY AS DETERMINATION

In an age when large corporations, such as Google, attempt to track our movements down to the meter and when data clearinghouses, like Acxiom, allegedly hold over 15,000 individual data points on each of our financial histories, the disruptive ring of Walter Benjamin's landline telephone comes across as quaint. Nevertheless, his comments on the culture of technologies connect past and present fears and hopes about technologies and their roles in society. After all, Hollerith's counting machine was both the foundation on which IBM was built and a key tool of the German Nazi Party's brutal eugenics policies during the Holocaust. In the 1920s, it also served as the basis for statistical applications in social work from which spring the racist recidivism algorithms that classify imprisoned people today. In tracing these histories, Brian Jefferson (2020, 78) deftly suggests Marx's analysis of industry can aid in the analysis of twenty-first-century digital Satanic Mills.[2]

The strengths of Karl Marx's analysis of technology are its direct connections to society and historical contingency, as we see at IBM. He approaches technology not as a primary focus, but as a contributing factor in his larger project, a political economic theory that demonstrates the inherent contradictions of capitalism. To him, technology acted as a sort of guiding precondition within a historical era, setting the stage for social relations at that time. In *The Poverty of Philosophy*, Marx (1955[1847]) writes "[t]he hand-mill presupposes a different division of labour from the steam-mill," a phrase perhaps best known in its aphorism form: "The hand-mill gives you a society with the feudal lord; the steam-mill, society with the industrial capitalist" (Marx and Engels, in Thompson 2008[1978], 201).

In the writings of subsequent orthodox Marxists, this was simplified into a technologically determined concept of society in which technology determines the type of economy (and class conflicts), and the economy outputs the culture. Such a simplistic approach misses the expansive, historically contextualized work of Marx himself.[3] In Marx's own immanent, revolutionary vision, competition forces continual revolution in the "instruments of production, and thereby the relations of production, and with them the whole relations of society" (Marx and Engels 1978[1848], 476). In other words, in part facilitated by technological innovation, the internal contradictions of capitalism move inexorably forward towards the inevitable victory of the proletariat and a classless society (Marx 1975[1852]; Friedman 1986).

While the hand-mill is a handy heuristic for launching a Marxist understanding of historical development and technology, Max Weber, the German social theorist and one of the founders of modern sociology, saw Marx's approach to technology as "simply wrong" (Weber 2005[1910], 26). In his early habilitation thesis work on class formation in the Roman era, Weber illustrates that "the same technology does not always denote the same economy, nor is the reverse always the case" (Weber 2005[1910], 27).[4] While Weber takes aim at the hand-mill aphorism, he is not dismissing Marx's political economic analysis, but instead a specific technologically determinist reading of economy and culture in Marx. For Weber, there is never a resting place that determines events in the ultimate last instance, instead emphasizing the dynamic movement of Marxian theory in which "everything relates to everything else" (Harvey 1999, xxix).[5]

In *Science as a Vocation*, Weber (1946[1919]) levels what's often considered his most trenchant critique of Marxism's view of technology as the arbiter of change. There, he posits the "alienation of reason from values, and its confinement to the instrumentality of bureaucracy and the aestheticism of a contemplative science" (Friedman 1986, 187). In such a condition, the inevitable victory of the proletarian remains trapped within the iron cage of capitalist modernity wherein the "technical and economic conditions of machine production ... determine the lives of all individuals who are born into this mechanism" (Weber 2005[1930],

123).[6] For Weber, seizure of the means of production is insufficient to escape modernity's iron cage as it was instantiated and reinforced through multiple registers of existence (such as religion), not simply the economic base of production. Further, the very role of science and philosophy, of thinking, was *not* to change the world as Marx had suggested in the *Theses on Feuerbach* (1978[1888]), but to contemplate it.

This clash of the possibilities and purposes of knowledge still echoes today at the intersection of science, politics, and daily life. Marx's struggle for social change versus Weber's contemplation of his iron cage continues as one aspect of the axis of hope and fear today. Concerns and hopes for technology are far older than the branding of the latest widget or revelations about how it violates users' privacy. However, these ideas operate at broad societal scales, and thereby miss the nuances of data and technology in actual practice.

A MENACING ALARM

Walter Benjamin inherited Marx's conception of distinct historical eras dominated by particular modes of production, and his work reflects both the hope and the fear of Marx's and Weber's discourses on tech-

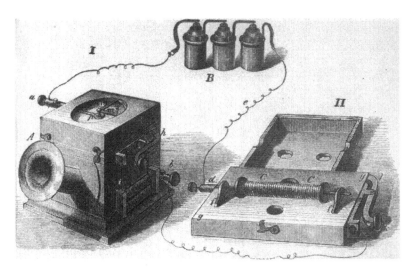

Figure 1.2   An 1863 image of Philip Reis' telephone from the German newspaper *Die Gartenlaube*. While Bell's legal patent eventually won primacy, Reis had invented a telephone-like device (and coined the term *Telephon*) in Germany by 1861. (Public domain)

nology. But Benjamin focuses not on modes of production and broad social forces, but inwards upon his experiences and rhythms of everyday life within those contexts. Born at the end of the nineteenth century, he is known for his work in Berlin and Paris in the 1920s and 1930s before perishing on the French–Spanish border while fleeing the Nazis in 1940. During his life, he worked prolifically to excavate how the dialectic conditions of production had seeped from the economic base into the cultural superstructure. This occurred in large part through technologies that had altered the way time and private space were experienced, thereby setting the very limits of what could be known amidst the bustle of daily life.

For Benjamin, so-called "Citizen King" Louis Philippe's ascension in 1830 marks the beginning of an era. In this period, "the private individual makes his entrance on the stage of history" and with them "living space becomes, for the first time, antithetical to the place of work" (Benjamin 1978, 154).[7] The home becomes a refuge from both commercial and social obligations, a private universe in which the individual may sustain themselves. This interior is the space that is disrupted and transformed decades later by the "violence" of the telephone (Benjamin 2008[1938], 78). It pierces this sanctuary of privacy, allowing the outside to reach into the home; to disrupt any quiet, any respite, with its clarion call.

Writing on technologically produced media and art, Benjamin asked two entwined questions of the transformations happening during his day (Jennings 2008, 9). First, what can art encode of the world around us? What can it reveal of the current epoch that would otherwise remain inaccessible and unknown? Second, how do modern media affect the human sensory apparatus? These ideas continue to resound today. For example, Benjamin examines the role of the camera as a tool of knowledge discovery (through slow motion, zoom, and so on) in shifting cultural expectations of truth and knowledge. These ideas prefigure arguments among Science and Technology Studies scholars about the roles tools play in the production and acceptance of knowledge and truth by decades. In *We Have Never Been Modern*, Bruno Latour (1993) sets a disagreement between Robert Boyle and Thomas Hobbes as a fundamental moment in the construction of modernity. He argues that Boyle's development of a process by which observations are made in a controlled laboratory (in that case, the use of vacuum tubes to manipulate and study the property of objects) contrasted with Hobbes naturalistic refusal. This clash "invents" the modern world, "a world in which the representation

of things through the intermediary of the laboratory is forever dissoci-
ated from the representation of citizens through the intermediary of the
social contract" (Latour 1993, 27). However, where Latour sees a separation of political and scientific power marked by the walls of a laboratory,
Benjamin sees the new technologies of his era as opportunities both for
disruption and the formation of new political and social solidarities.

Film, for Benjamin, is the foremost dissecting, dehumanizing technical
apparatus, as it literally breaks life down into a series of still, jerky images.
In this, it represents the subjugation of life to the assembly line in both its
production and consumption (Benjamin 2008[1935b], 340). At the same
time, once the camera is capable of making new discoveries, its very dissecting nature allows it to perfectly reproduce humanity against capitalist
modernity, a "vast apparatus whose role in their lives is expanding almost
daily" (Benjamin 2008[1936], 26). The camera both offers the ability to
reproduce one's humanity and simultaneously extends with whom this
experience of humanity may be shared (Benjamin 2008[1935b], 100).
Without using today's buzzwords, Benjamin argues that film offers the
first of what we would now call social media. It grants the masses the
ability to share their humanity across time and space in ways previously
unimaginable. In a moment of hope, he suggests it offers a pathway to
liberation as the proletarian mass will be able to understand "themselves
and therefore their class" (Benjamin 2008[1936], 24).

What Max Weber had challenged as impossible, creating art "com-
prehensible to other members of his class" (Weber 2005[1910], 28),
Benjamin sees as feasible through the incipient social media of popular
cinema.[8] In fully reproducing class identity and fostering class solidarity, film was a means of politicizing art, of creating an aesthetic register
capable of directly invoking revolutionary action. It is a means of reas-
serting shared humanity in the face of a technological apparatus that
seeks to order, calculate, and control. If we follow this line of thinking
with respect to the spatial data we produce every day, we can see that
data come to *stand for* individuals in ways both banal, such as targeted
ads for new shades of lipstick, and violent, such as the targeting of drone
strikes based upon phone metadata. But we also see that data might also
open spaces and moments for new joyful encounters and alternative
political economic relations. The multiple axes of hope and fear exist
simultaneously in dialectic tension with one another. Benjamin was not
and is not alone in these ideas, and tracing their subsequent development better contextualizes the current political economy of data.

THERE IS NO ALTERNATIVE

I bought a bourgeois house in the Hollywood hills
With a truckload of hundred thousand dollar bills.
Man came by to hook up my cable TV.
We settled in for the night my baby and me.
We switched 'round and 'round 'til half-past dawn.
There was 57 channels and nothin' on.

> (Bruce Springsteen, "57 Channels (and Nothin' On),"
> *Human Touch*, 1992)

Publishing after Benjamin's death and the defeat of fascism in Germany, Horkheimer and Adorno take up the question of capitalist society's "irrational rationality" in the face of Stalin's authoritarian Soviet Union and the post-war economic boom of the United States. Their view of technology is not wholly dissimilar from Benjamin's: technology has seeped ever more into our lives, which have become dominated by a "unanimous" system of "aesthetic manifestation of political opposites [that] proclaim the same inflexible rhythm" (Horkheimer and Adorno 2002[1947], 94). However, where Benjamin took solace in the radical art appearing in the Soviet Union, its disappearance late in Stalin's reign signaled the Soviets' desire to "turn the world into an enormous workhouse" (Horkheimer, in Friedman 1986, 191). For Horkheimer (1995), his work on the failures of the Enlightenment and attempts to reclaim science with a critical spirit were a response to the perceived collapse of "critical theory" in the Soviet Union at the time. In their bleak "relentless complaint about technology and techno-culture," Donna Haraway finds a direct engagement with Weber's iron cage of life in light of its seeming confirmation in post-revolutionary Soviet society (Gane 2006).

A key tenet of Marxist thought is the ways in which capitalists reduce labor from qualitative differences (*what* one does and makes, *who* one is) into the quantitative differences of their outputs (*how many* items they make).[9] Agreeing with Benjamin, Adorno and Horkheimer see a similar process playing out in the "non-productive" or reproductive areas of society, the "superstructure" in Marxist terminology. They argued that the "culture industry" had created a social hierarchy that ranked humans as consumers placed into different *calculable* brackets.[10] The quantitative ranking for the purposes of consumption created a false sense of choice within a society, allowing individuals to develop fierce loyalty to certain brands, such as Ford versus Chevy trucks. Thus, making a consumptive choice becomes based on differences so illusory that they can be recognized "by any child" (Horkheimer and Adorno 2002[1947], 97). In this way, humans become reified even to themselves, judging value by objects owned and brand loyalty. In a world where personal brand worth is measured by retweets and Instagram followers, it is not difficult to see immediate, deep connections between this insight from the mid-twentieth century and our present moment.

This reduction to rhythmic, repeatable quantification occurs at the levels of both consumption and the ideologies that enforce it. In Adorno and Horkheimer's argument, science and philosophy have lost their critical edge, descending into a quantitative practice of measurable outputs that describe and classify. For example there is an almost feral quality to citation counting within academia, and at times a poor, biased, or even fabricated publication may accrue thousands of citations because it makes for a convenient punching-bag.[11] Instead of thinking and building alternatives to the present, research retreats into an ivory tower in which its most valorized practitioners decree that values and norms of present-day society always have and always should remain the same in perpetuity. "Well, first of all, tell me, is there some society you know that doesn't run on greed?", as the Nobel prize winner Milton Friedman put it to Phil Donahue in a 1976 interview, cavalierly ignoring centuries of work in history, anthropology, and other fields in the name of orthodox, twentieth-century economics (Donahue 1976).

Where Benjamin saw hope in emerging technologies and new forms of art as opportunities to realize our shared class consciousness and reassert our humanity, Horkheimer and Adorno found only the "sardonic realization of man's species-being." They saw a consumptive, flattened ideology that insists that the present is inevitable, natural, and unalterable, "hid[ing] itself in probability calculations" (Horkheimer and Adorno 2002[1947], 116). In their line of thinking, the ideological reduction of possibilities, and the coinciding retreat of science into its ivory tower, controls the inherent tensions and contradictions found within capitalism. Where Marx sees the contradictions within capitalism's productive economic base as steps towards an inevitable systemic collapse and social liberation, Benjamin, Horkheimer, and Adorno find little solace. They see that these contradictions have seeped into (or always-already have been, as Max Weber previously observed) the ideological underpinnings of society, and Horkheimer and Adorno find the contradictions to have been stabilized through management.[12] The culture industry sets the stage for humans to accept, and even identify with and revel in, the banal choices presented to them. There is no outside alternative, but a million variations on what brand of jeans to wear. Benjamin's fervent hope for technology to realize class consciousness in the face of fascism has too often become an epistemological cry of despair in the consumerist post-war world. Simply from a historical perspective, Benjamin was proven wrong and ultimately lost his life for it, while Horkheimer and Adorno have not yet been. "Someone once said that it is easier to imagine the end of the world than to imagine the end of capitalism" (Jameson 2003, 76). The tensions and contradictions favoring social change may have risen, but so too have the forces aligned to negate and manage them, from geodemographics to individually targeted advertising, from closed-circuit television (CCTV) cameras in stores to street corner cameras tied into facial recognition systems. So why talk about Benjamin?

WAYS OUT?

The purpose of this book is to use critical thought to inform daily practice in ways that reshape and repurpose the data that have come to stand for us in systems of capital, surveillance, and governance[13] in ways that produce hopeful, productive moments of joy and solidarity. Such possibilities are at the heart of Benjamin's *The Work of Art in the Age of Its Technological Reproducibility* (2008[1936]). The power of technologically enabled creativity in the right context is the potential for change and new ways of life:

> Whether it is a self-taught engineer designing machines using scraps scavenged from junkyards, a performer developing a new instrument using electronics and open-source technology, a comic book artist creating an imaginary world that uses secret technologies, or an architect using aesthetics to bring about social change, [techno-vernacular creativity] practitioners sample (simulate), reappropriate, and remix heritage artifacts and technologies to generate works that can be embedded into different environments.
>
> (Gaskins 2019, 272)

We contend that critical thought can help in understanding the technological apparatuses which seek to manage and dominate our lives. Moreover, critical thought's conceptual tools can assist in identifying and realizing contextual technological openings and possibilities that can facilitate more empowered, humane, equitable relationships and ways of living through technology.

Walter Benjamin joins a long list of optimists, from John Dewey to Marshall McLuhan to the self-serving promises of Mark Zuckerberg, whose hopes have been dashed when technology does not inherently deliver better ways of life, much less some kind of utopia. Instead of realizing better ways of life, Zuckerberg's personal net worth exploded. Yet for all the trials of the last century, there remains a narrow glimmer in Benjamin's work. If one reads his focus on creativity for social change not as inherent or predetermined outcomes, but instead as possibilities that could be leveraged and realized with dedicated effort, alternatives can be found. With that approach, technologies can, with care, become tools for helping produce solidarities and better daily lives.

As we've already seen in this chapter, technologies do not determine society, nor are the outcomes of a technology wholly up to the individuals using it. Technologies are socially, historically contingent; no technology, from data to algorithm, can fully escape the material conditions in which it is both designed and used. Current contexts of digital data, systems of counting, of ordering and sorting continue to spiral into new potentially productive and ideological terrain. Applications rate everything from our Facebook photos to our creditworthiness to our potential sexual partners. In light of the material failure of Benjamin's hopes, we consider three further potential routes of escape, ways to approach those possibilities for human life amidst, with, and through technology: one abhorrent (Heidegger), one provocative (Debord), and one earnest (Marcuse). These three are hardly the only thinkers to critically engage this question. Others have come from other spaces, places, and orientations, as we will continue to highlight throughout this work. However, interweaving threads from these particular three allows us to draw forward and form the basis for productive radical practices of resistance.

ONLY A GOD

Martin Heidegger was a Nazi. While his public persona always attempted to eschew this reality after the war, publication of his so-called "black notebooks" after his death has removed any pretense or doubt.[14] Nevertheless, his thinking on the relations between technology and society and his pursuit of alternatives and escapes from the excessive alienation of humanity under modern capitalism cannot be so easily dismissed, especially in light of our actually existing world.

Heidegger makes clear that technology is in no way neutral or impartial. First, viewing it as such delivers humanity "to it [technology] in the worst possible way" (Heidegger 1977b, 4), akin to Thoreau's line that people "have become the tools of their tools" (Thoreau 1910, 47). Second, while "[s]cience is the theory of the real," the real will always approach mankind through modern technology (Heidegger 1977b, 157). Third, "[t]echnology is ... no mere means. Technology is a way of revealing" (Heidegger 1977b, 12). These three points concerning the neutrality of technology, its relationship to the real, and the revealing of technology provide powerful insight into the means and limits of technology as it functions in our world today.

Heidegger's insights build upon the phenomenological movement in philosophy. In this line of thinking, "things" are objects that appear as present in consciousness as "things themselves" without having a hidden or deeper existence (Husserl 2001). Heidegger's thinking insists that as a technological object functions, it retreats from conscious consideration and becomes a "thing" taken for granted (Harman 2010). In modern terms, we rarely notice when our phones seamlessly find a restaurant nearby when we are looking for dinner in an unfamiliar place. Only a few years ago, choosing our cultural and culinary experience would have been up to a generic guidebook, a friendly recommendation, or chance, not a GPS-indexed, personalized algorithmic ranking. Digital media scholar David M. Berry has highlighted modern technology's particular ability to elide the technical considerations by which it works. He suggests that algorithms have moved people from *knowing that* to *knowing how* (Berry 2011, 121). We know *how* to search for a restaurant using Yelp, but we do not stop to think about the obfuscated system of hardware (processing units, cell towers, and so on), software (operating systems, Yelp's app), and data (Yelp's individualized profile of each of us) that make finding a restaurant possible.

Heidegger's insistence that technology is not and cannot be neutral helps dispel the liberal belief that "it's not the technology, but how an individual uses it."[15] The idea that a technology itself, regardless of the intention of its users, may have intrinsic orientations towards the world is a powerful one. It undercuts the very structures on which narratives of tech solutionism are built. Take, for example, the ousting of Timnit Gebru, the former co-lead of Google's ethical artificial intelligence (AI) research team. In short, Gebru and her co-authors suggested that there were underlying, structural, and even intractable risks associated with training artificial intelligences on large language models. For all the seemingly meaningful texts their AIs produced, they suggested real, material dangers lurked both in the environmental impacts of running such models (and the resources they consume) and also in a shift from *understanding* to *manipulating* language ("knowing that" to "knowing how"). These problems, Gebru and her co-authors suggest, will remain regardless of steps taken to mitigate them and regardless of how profitable their application may be. Google's management, citing disagreements about their conclusions, which would coincidentally impact the company's bottom line, dismissed Gebru.[16]

Mobile spatial applications and the data they produce, extract, analyze and help users enact both performative and normative tasks: apps announce locations at specific events to a users' friends (performing our mobility), they recommend which meals to order at a restaurant (normalizing what is and isn't eaten), and they suggest nearby potential romantic partners. As technology comes to mediate and shape these actions, according to Heideggerian thinking, everyday life transforms into a condition of *standing-reserve* (Berry 2011; Heidegger 1977b). This is an important shift in the orientation of life and is central to Heideggerian thought on technology and potential escapes thereof. *Standing-reserve* can be thought of as a counted stock. It is something that awaits its own use in an orderly, known, manner (Edwards 2007). Take, for example, an airliner waiting on a runway. It reveals itself as *standing-reserve* in that it is "ordered to ensure the possibility of transportation"(Heidegger 1977b, 16). Any deeper meaning to its function and purpose is left unconsidered: there is a plane, it will fly some number of people or things to some other location.

At their most bombastic, tech evangelists see the world as *standing-reserve*. In an infamous *WIRED* piece from 2008, then editor-in-chief Chris Anderson wrote of how big data signaled the "end of theory," a situation in which it didn't matter *why* things happened, only that they could be successfully predicted and modeled.[17] In this formulation, the fullness of the world can and will be explained entirely through quantified, calculated sets of information, the world as *standing-reserve*. Spatial data are particularly useful in such a rendering as it attempts to reduce the fullness of a person's experiences and observations, in Heideggerian terms their *being-in-the-world*, into a set of quantified coordinates, latitude, longitude, time, and so on.[18] These can be linked to other data and indexed by location to call forth consumptive practices. The leap from traditional geodemographics, aimed and based upon one's home postal code, to one's bodily location is an attempt to move towards a smoother, more predictable, quantifiable space of capitalist consumption. The so-called "killer app" of the twenty-first century has been described as the ability to guide a consumer from their every waking moment to their next purchase (Krumm 2011). As Anderson's triumphalist call makes clear, this is done not through any deeper understanding of the motivations of said consumer, but through the processes of creation, extraction, interconnection, and analysis that make up so-called big data.

Heidegger's insight into life as *standing-reserve* and the global spread of calculation and quantification demanded by modern technology have clear implications for how we make use of and are used by our everyday technologies. However, his response to these conditions marks a distinct retreat from any sort of practical engagement with the world. Comparing Marxist thinking with Heidegger's thought, geographer Stuart Elden (2004) observers that, for Heidegger, the relations between society and technology set the very conditions for the possibility of science in the modern world: all thinking occurs within a specific *enframing* that sets the limits of what may be recognized and researched. Within capitalist modernity, where technology has come to demand life as *standing-reserve*, this *enframing* is one in which philosophy has been taken over by a science that is overdetermined by what can be known of and through technology (Heidegger 1981[1976]; Harman 2010). Science, for Heidegger, is an endeavor which "entraps the real and secures it in its objectness" (Heidegger 1977a, 168). It is a means by which *being-in-the-world* is reduced to *standing-reserve*.

In this line of thinking, philosophy has become unable to change the world, and any belief that it may is an example of the worst possible kind of arrogance of humankind (Heidegger 1981[1976]). By the end of his life, the only possible way out that Heidegger sees is a removal of humanity as the object of history, allowing for things "to be" in a way not mediated by human goals and desires: *only a god can save us* (Heidegger 1981[1976]). While his point that technology is inherently not neutral is valuable, Heidegger's conclusion is paralyzing. Rather than waiting for divine intervention, others in the post-war period attempt to take technology, as non-neutral as it is, in more productive directions.

AN OFF RAMP

Even as Heidegger published his work on technology, members of the soon-to-be Situationist International were beginning to develop a radically different approach to imagining ways out through activist politics and culture jamming. Emerging from Marxist thought in post-war Paris, the Situationists, chief among them Guy Debord, approach technology as directly tied to capitalist production. However, they break with traditional Marxist thought by emphasizing culture and technological media as a form of production. Technological media aren't limited to capital, for they also include the labors of people going about their daily lives. In practice, they attempted to utilize these ideas in a wide variety of arenas, from urban planning to film to acting as key instigators in the May 1968 Revolution in France.

One of the key themes running through Debord's work is how capitalism attempts to colonize and extract value from the practices of people's everyday lives. Doing so involves stripping a person of their authentic life and replacing it with spectacle, commodified images of such a life: "All that once was directly lived has become mere representation" (Debord 1967, Thesis 1). He continues:

> The images detached from every aspect of life merge into a common stream in which the unity of that life can no longer be recovered. *Fragmented* views of reality regroup themselves into a new unity as a *separate pseudo-world* that can only be looked at. … The spectacle is a concrete inversion of life, an autonomous movement of the nonliving.
> (Debord 1967, Thesis 2; original emphasis)

This technological separation and replacement with commodified images and identities is apparent not only in film and commercials, but also the places we frequent. Billboards and product placement fill our visions, separating ourselves from real experiences and replacing them with imagery of social interactions and desires for purchase.

The concept of the spectacle directly connects capitalism's need for growth with everyday experiences as capital's "corporeal corkscrewing inwards" colonizes more and more everyday, banal experiences and interactions and integrates them into market exchange (Beller 2012, 8). This is a totalizing concept, "the autocratic reign of the market economy which had acceded to an irresponsible sovereignty, and the totality of new techniques of government which accompanied this reign" (Debord 1998, 2). Some argue this limits its utility, for how can we resist that which is total? Is it not better to, as Heidegger suggested, wait for a god? But, as we've written elsewhere (Thatcher and Dalton 2017, 137), such views mistake "a totalizing tendency for a static totality." The point of the spectacle is that it dynamically attempts to colonize and subjugate daily life at all levels of experience and at all times. This dynamic tendency opens fertile ground for resistance, both among the Situationists and later activists who take inspiration from them.

The Situationists are not known for direct success in their resistances, whether overthrowing capitalism or shaping the mid-century redevelopment of Paris. Rather, as we see with the May 1968 Revolution, their ultimate impact tends to be cultural, which reflects the nature of their methods. They developed and practiced their own resistance methods, such as détournement, artistic repurposing of existing advertising and spectacle imagery turned back on itself. Perhaps most relevant to geographic data is the dérive (drifting), a form of semi-systematic wandering to better know and articulate a place.[19] Such cultural practices are significant because the Situationists see the possibility of a world built on play, an alternative to the howling feedback loops of consumption in which capitalism entrenches us. Moreover, such play opens paths through engagements with and repurposing of existing systems, destruction through plagiarism of ideas, media, and actions (Wark 2013). A similar ethos suffuses this work.

EARNEST MULTIDIMENSIONALITY

If Heidegger turns to fascism to displace the iron cage of modernity, and when that fails, waits in faith, and if Debord insists on playing the role of avant-garde urban provocateur, then Herbert Marcuse offers our third pole for (re)asserting our shared humanity in the face of the current sociotechnical apparatus: not through listless waiting or art-fueled subversion, but through a radical synthesis of Heideggerian and Marxist thinking. Marcuse re-orients Heidegger's *enframing* to encapsulate capitalism such that he offers a productive engagement with technology amidst capitalist modernity.

Marcuse studied under Heidegger at the University of Freiburg, only to later reinvent himself as an engaged Marxist social theorist. While it was Heidegger's early support for National Socialism that led the two men to break both politically and personally, Marcuse's thought on technology and scientific reason retain Heidegger's influence (Thomson 2000).[20] For example, Marcuse rejects the neutrality of technology due to social context. In the modern world, there is "no personal escape from the apparatus which has mechanized and standardized the world" (Marcuse 1982, 143). Similarly, science functions under a "*technological a priori* which projects nature as potential instrumentality, stuff of control and organization" (Marcuse 1991[1964], 153). Although he eschews the terms *standing-reserve* and *enframing*, the underlying ideas remain.

However, Marcuse combines this with a Marxist rethinking of technology as part of a *mode of production* within a larger socioeconomic system of capitalism. For him, technology represents a "mode of organizing and perpetuating (or changing) social relationships, a manifestation of prevalent thought and behavior patterns, and an instrument for control and domination" (Marcuse 2009, 138). And here, Marcuse makes a most critical intellectual move: technology is all of those things as well as potentially, and necessarily, a specific technique of liberation. That which surrounds society, that which renders it flat and hides the depth of true being (the *enframing* for Heidegger), *is* capitalist modernity for Marcuse. Technological rationality calls for nature in a calculable and orderable manner *and* creates a standardized excess of resources and humans ready to be called forth through technology. But in so doing, "technological rationality has become political rationality" in which a totalizing system of capital has emerged that is "everywhere and in all forms" (Marcuse 1991[1964], xlviii, 11).

To avoid Heidegger's dead end, Marcuse inverts both Heidegger and Weber by suggesting that it is the very commitment to truth, to "real facts," that instills the scientific endeavor with its critical spirit, with its desire and need to change, not just interpret, the world (Marcuse 2009; Friedman 1986). Where Weber saw this as utopian fantasy, Marcuse reads a political necessity, a charge where "[t]ruth becomes critique and accusation, and accusation becomes the function of true science" (Marcuse 2009, 161).

Echoing Heidegger and Benjamin, Marcuse wrote of existing as an instrument, as a thing that could be called forth to function in predictable, expected ways. Here was a "society without opposition" that had grown directly out of "technology as a form of social control and domination." But rather than effectively giving up on human agency, as Weber, Adorno and Horkheimer, and Heidegger had, Marcuse harkens once more to Benjamin, ending his work declaring "Nur um der Hoffnungslosen willen ist uns die Hoffnung gegeben" ("It is only for the sake of those without hope that hope is given to us"; Marcuse 1991[1964], 257).

Forty years after his death, Marcuse's work remains more pertinent than ever in its lucid insistence both on the role technology plays in society and in the need and possibility for alternatives. Our winding path from neoliberal myths of technological triumphalism back through the mills of Marx and Weber and into Heidegger's contemplative retreat have all led us here: Marcuse matters because of his insistence that alternatives must not just be thought, but also acted.

That is the core concern of this book: it is absolutely necessary to both think through and act to produce alternatives. Radical thought and practice form a nexus through which alternative technologies can emerge. In this we echo and extend calls that we trace back through Benjamin, and that reverberate through academic and some popular presses even as our relations with and representations in spatial data grow apace. In his beautiful work *All Data Are Local*, Yanni Alexander Loukissas (2019, 162) develops an idea of *critical reflection* as "a process by which the interwoven social and technical dynamics of data are made visible and accessible to judgement." If that constitutes reflection, then what does it mean to think and enact a radical praxis of spatial data, to reorder our relations with technology? In the final section of this chapter, we present two additional, recent examples of spatial data and the knowledges they enable bursting beyond their intended uses and slipping into the public view. Thinking through these approaches using the reflection of Heidegger, the playfulness of Debord, and the inversion of thought and action provided by Marcuse allows us to move from the depths of social theory in the mid-twentieth century into our daily routines today.

THE CURIOUS CASES OF STRAVA, AND THE HOME OFFICE'S CHARITY MAP

According to McKinsey, a well-known management consulting firm, the number of Internet of Things (IoT) (and thus at least passively location-aware) devices will surpass 43 billion in 2023. That would constitute a €14.5 billion worldwide market (Dahlqvist and Patel 2019). While the scale of this may be surprising, the normalization of quantified self-tracking is likely not new to many readers. We track our packages, our runs, our diets, our sleep patterns. At larger scales, corporations track our purchases, our web views, our likes. Digital personal assistants sit on our tables and in our pockets listening to our conversations, and governments are well established to be tracking not only traffic patterns, but also our private conversations and exchanges.

Much as the ways of thinking through the relations between society and technology are not new, nor are the practices that prompt the necessity of such thinking. Benjamin Franklin enjoyed self-quantification through personal charts of how he spent his time and whether he lived virtuously or not (Neff and Nafus 2016, 15) while James Bridle relates the story of Robert Lawson who, in 1967, revealed that the UK government was collecting a copy of every telegram that entered or left its nation (Bridle 2018, 175). We have lived for some time in what geographer Matt Wilson described as a "quantified self-city-nation"—a space in which:

> the flickering screens, the dynamics of real-time data and the prospect of behavioural change intersect in a glossy imaginary where being technologically fashionable and facile supersedes concerns of differential docility. ... [W]e are assured of the untapped potential at the touch of the flat screens in some of our pockets, that the possibility of our "fittest" bodies and "smartest" cities rests with individual behaviour.
>
> (Wilson 2015a, 39)

In post-War America, Horkheimer and Adorno (2002[1947], 71) decried a reduction of personality to "hardly more than dazzling white teeth and freedom from body odor and emotions." Now we are promised something both shinier and more sinister. In the quantified self-city-nation, the data we generate, the steps we take, our swipes to the left, the rides we share become our outward-facing personality. This representation is algorithmically produced through the analysis of said data that comes to stand for us both in systems of global capital (our mortgage rate, what jobs we are offered) and our personal, private moments (potential partners offered through an app, the directions we are given through a city). While this quantified reduction to data is sold as our best selves, as a life easily and fulfillingly lived, *fitness at ten thousand (counted) steps a day,* it masks a deeper slippage between who we are in the world and who our data say we are. When these two forces rub against one another, they can produce unexpected results.

Founded in 2009 by Mark Gainey, Mark Shaw, and Michael Horvath (all white men), San Francisco-based Strava bills itself as the "#1 app for runners and cyclists" and raised over $110 million in funding in 2020 alone to support this goal (Etherington 2020). You may have used it or something like it, such as RunKeeper, MapMyRun, Fitbit, or Garmin. In essence, it's an app that tracks your exercise using the GPS sensor in your smartphone. Even if you have not used such an application, you've likely seen someone doing so: checking their phone or smartwatch at the beginning or end of a run or sharing a particularly good time on social media. This is the sort of banal practice through which quantification and tracking seeps into our everyday.

In November of 2017, Strava created a visualization of all their users' paths in the form of its Global Heatmap. Made up of 5 terabytes of input

Figure 1.3   Images at two scales around Boston from the "One Dot Per Person for the Entire United States" visualization created by the Demographics Research Group at the University of Virginia. A dot-density map built upon the 2010 census enumeration. Though blurry here, this style is emblematic of many current approaches to data visualization. At regional scales (the left), the dots blend into one another, producing what appear to be filled areal units; however, when zoomed in to more local scales (the right), the dots disaggregate and show the emptiness within said units. (Image copyright © 2013 Weldon Cooper Center for Public Service, Rector and Visitors of the University of Virginia, Dustin A. Cable, creator)

data and 1.4 trillion coordinate points, it promised the "largest, richest, and most beautiful dataset of its kind" (Robb 2017). Why build such a map, on a worldwide scale so far divorced from the distance that people actually run daily? Chief Technologist of User Experience and Sustainability at Intel Melissa Gregg points to a specific scopophilic pleasure to be found in viewing large data sets. Building on Orit Halpern's *Beautiful Data* (2014), she describes the appearance of "command and control through seeing" that follows from the rendering of massive sets of data imaginable through their visualization (Gregg 2015). Figure 1.3 illustrates how in such over-loaded visualizations, data points bleed together at more extensive, regional scales, only to pull apart at more local scales, providing the illusion of analytical meaning. To see is to know, to imagine the whole of a complex system that constitutes data points beyond the scale of which we could otherwise think.

Nonetheless, visualizing this kind of spatial information, connecting who, when, and where, has profound implications beyond aesthetics, as quickly became apparent in Strava's Global Heatmap. On January 27, 2018, Nathan Ruser, then an analyst for United Conflict Analysts, tweeted:

> Strava released their global heatmap. 13 trillion GPS points[21] from their users (turning off data sharing is an option). https://medium.com/strava-engineering/the-global-heatmap-now-6x-hotter-23fc01d301de … It looks very pretty, but not amazing for Op-Sec. US Bases are clearly identifiable and mappable[22]

This marked the beginning of a Twitter thread that explored examples of Strava's Global Heatmap making visible various military bases and patrols around the world. Within days, the story had been picked up by *The Guardian*, the *BBC*, *WIRED*, and a host of other online, print, and television media venues (see, inter alia, Hern 2018; BBC News 2018; Hsu 2018). Strava, in response, made a number of changes both to who could view the Global Heatmap (registered users) and how its data were stored (data marked private was deleted monthly). Moreover, a second wave of articles in the news media described how to better protect your data when using such applications.

In the case of the Strava Global Heatmap, what was put at risk were the secrets of some of the best-funded and most powerful organizations in the world, the US military and its affiliates. Nonetheless, this type of

Figure 1.4 A clearly demarcated US military base discovered in Strava's Global Heatmap by Nathan Ruser. (Image used with author's permission)

unintended visibility through spatial data production and its consequences also occurs at other scales with more vulnerable populations. One such case was the UK Home Office's use of data from a London homelessness charity to identify and deport homeless people from the country.[23]

The story, extensively covered in *The Guardian*, began when local charities began to collect data from the homeless people they worked with, including nationality and where that person frequently spent time on the streets. With the charities' consent, the Home Office gained access to this information and used it to identify, locate, and deport non-native individuals. When this secret partnership became public knowledge, public outrage and a government investigation ensued. As we wrote with Clancy Wilmott and Emma Fraser in 2020 (Dalton et al. 2020):

The data in question was part of the Combined Homelessness And Information Network (CHAIN), a shared database funded by the London Mayor's office and administered by St. Mungo's, a major homelessness charity, with access to the data granted by the Greater London Authority (GLA). Each homeless person has a listing, logged by charity outreach workers, with their name, history of homelessness, special needs, gender, age, and crucially, their regular location and nationality …. In May 2015, the Home Office secretly obtained GLA permission to access the CHAIN database. That data now served a Home Office program to remove homeless non-UK nationals. If a homeless person declined contact or refused an outreach worker's offer of help to voluntarily return to their country of origin, the Home Office would send officers to their regular location to detain and deport the person by force …. CHAIN geographic data facilitated the detentions by indicating where to find the homeless person in question, and, in the aggregate, geographic "hotspots" where potential deportees might be concentrated. "We are trying to build in a timeline on the map so you can see where non-UK nationals have moved to over time, which hopefully will also be able to help you establish priorities by seeing patterns …" [an officer from the Home Office told the press] … Deportation rates rose an estimated 41% for EU nationals, totalling 698 EU nationals deported by May 2017.[24]

In both the case of Strava and the case of London homelessness, we find the representational ability of spatial data slipping beyond the intent of their creation. In both cases, lived experiences, where one runs, where one sleeps, become translated into aesthetic representations of geographic "hotness," where one is likely to be/have been at some time. These two disparate registers, the state military secret and a homeless sleeper, are united in a visual rhetoric which promises calculation, control, and predictability through the mapping and visualization of spatial data. And as we see, that visual rhetoric and the data behind it easily slip beyond the original intent of data creators.

This kind of control based on population is the logic of the carceral state. It is clearly visible not only in Strava or the Home Office, but in the unfortunately popular genre of crime "hotspot" maps. The individual person necessary to commit an act is completely effaced, replaced by an areal representation of likelihood for *someone* to commit *some* crime within said area (Jefferson 2017).

Even if the topic isn't crime, data assembled into such maps become an aesthetic representation of "hotness," where one is likely to be/have been at some time, where some act has occurred or is likely to occur. In *New Dark Age*, James Bridle suggests that the very excess of information creates the very conditions in which it is unlikely for us to think and to know in detail; with every new piece added to the pile, the world expands beyond our ability to conceptualize and threatens our ability to act in it. Mapping and visualization offer a path through this fog. In the face of overwhelming data, Orit Halpern (2014, 22) illustrates how these methods become a "map for action" that transforms what it means to know through what it means to see. Pivoting from nineteenth-century motifs of extracting value from natural resources, she writes of how in the twentieth century there emerged:

> an aspiration and desire for data as the site of value to emerge from the seeming informational abundance once assumed to be the province of nature. Data ... appeals to our senses and can be seen, felt, and touched with seemingly no relationship to its content.
>
> (Halpern 2014, 15)

Large, location-indexed data promise such control through visualization, life as *standing-reserve*, placed, known, circulated, and called forth for algorithmically sorted purposes at algorithmically calculated times. This is part of data's promised leap from geodemographics to individualized marketing. The focus is no longer "people in this neighborhood are likely to buy trips to Cancun over the summer," but rather "Jim Thatcher will purchase a purple sweater this coming Tuesday." More data, through algorithmic processing and visualization, are transformed into a means of governance and consumption.

Jennifer Gabrys (2016, 41) writes of how "[p]rocesses of producing data are also processes of making sense." Drawing on Foucault's analysis of how neoliberal thinking permeates governance and its subjects, she suggests that efforts to make our environments more computational produces a "biopolitics 2.0," which:

> [e]merges within smart cities that involves the programming of environments and citizens for responsiveness and efficiency. Such programming is generative of political techniques for governing everyday ways of life, where urban processes, citizen engagements, and gover-

nance unfold through the spatial and temporal networks of sensors, algorithms, databases, and mobile platforms.

(Gabrys 2016, 203)

Quantification becomes representation at scale, allowing spatial data to become the individual that capital can see.[25] It is not the individual as they might understand or describe themselves, but rather an amalgamated population built around similar data points within a matrix according to the chosen algorithmically based analysis. To accomplish this, sensing apparatuses have shifted from discrete points, such as CCTV cameras and credit card machines, to what Greenfield (2006) described as "everywhere": our phones, our homes, even our cars and refrigerators can now provide a litany of information tied to where we are, when, and what we were (allegedly) doing. These data are not so much a reflection of who we are, but a construction of how capital and government see and attempt to manage us. Our UberEats orders are linked to our Tinder matches which are tied to our Facebook accounts that in turn advertise us products from our Amazon wishlists. Moreover, as social processes, the benefits and penalties are not equitably distributed. The carceral logic at work is clearly apparent in the racially biased outcomes of even "impartial" services and data connections (Benjamin 2019a).

Nevertheless, there remains a rupture, a crack that opens as data are transformed from their expected, intended uses into moments of representation. In practice, this excess and its unexpected outcomes remain a source of apparently constant surprise.

FROM GIRLS AROUND ME TO DRONES ABOVE ME AND BACK

In 1993, Oscar Gandy Jr. published *The Panoptic Sort: A Political Economy of Personal Information*, in which he attempted to outline the growing stakes of informational sorting *vis-à-vis* personal privacy and political autonomy (Gandy 1993a). A prescient and sadly out of print book, Gandy's fears (and hopes) seem all the more pertinent in light of Acxiom's supposed 15,000 interlinked points on every United States citizen and the National Security Agency's (NSA's) supposed recordings of all digital communications. He illustrates again and again the dangers of viewing personal information as a commodity:

In general, we wish to suggest that unlike the value generated when free labor enters into a contract with capital to sell its labor power, much of the personal information gathered by business and government is collected either surreptitiously, or under circumstances which reasonably can be understood as coercive.

(Gandy 1993b, 82)

Writing in 1995 about asymmetries of access to information, Eric Sheppard similarly asked, "What does it say about the influence of social power over information systems ... when individual credit card ratings are available to private firms, whereas detailed financial information about those firms are defined as proprietary information[?]" (Sheppard 1995, 12). The scale may change, but the intent does not. When discussing risk management through data analysis, Gandy noted how rental car agencies refuse service or charge more to local renters due to perceived risks of unsafe driving or theft, so that "[c]lass membership then predicts individual experience" (Gandy 1993b, 88). Today, it is increasingly the norm for rental agencies to use GPS trackers to monitor driving habits which, in turn, place renters into representative categories based on determined fares. The scale has changed, the intent has not.

Twenty years later, writing for *The New York Times' Bits* blog, Nick Bilton (2012) seemed fatigued by the latest invasion of privacy through spatial data: "Another day, another creepy mobile app. Here is one that allows you to find women in your area. It definitely wins the prize for too creepy." Bilton was describing the short-lived and infamous mobile application Girls Around Me. It linked together Facebook and Foursquare through their Application Programming Interfaces (APIs)[26] within the Girls Around Me app to provide a near-real-time profile of each of the women near the device. In many ways, it functioned like Tinder or Grindr do today. Girls Around Me would check nearby locations (typically bars and restaurants) for women who had checked in to that location using the Foursquare application. Then it would attempt to pull up their Facebook profiles, providing interests, likes, and photos. Of course, the core concern here is that while users intentionally create and log into Tinder, Girls Around Me did this by repurposing women's data generated by Foursquare and Facebook without the women's knowledge. Unlike Tinder, in which a user can choose what information to make visible to potential sexual partners, disclosing more information (such as actual location) at their discretion, Girls Around Me provided check-in locations and the full suite of information available publicly through their Facebook accounts without consent.

Once "discovered" by a blogger at *Cult of Mac* (Brownlee 2012), press coverage was generally outraged and appalled in tone. Only a few bloggers, such as Kashmir Hill of *Forbes* noted that the app functioned like certain dating applications, specifically Blendr, a forerunner of Tinder based upon the success of the gay-oriented Grindr application. She suggested: "I'm sorry, my friends, but I think apps like 'Girls Around Me' are the future" (Hill 2012). Hill also, quite correctly, noted the patriarchal tones in coverage of Girls Around Me in which women were framed as damsels in need of protection from a lack of understanding of security features. As she put it, "Sometimes we can be found because we want to be found" (Hill 2012).

About a year later, in June 2013, Edward Snowden set off another data scandal by releasing thousands of purloined NSA documents to a group of journalists at outlets like *The Guardian* and *The Washington Post*. The narratives and counter-narratives of Snowden and the documents he released are now legendary, having been presented in a variety of media forms, up to and including a dramatic biopic starring Jason Gordon-Levitt and directed by Oliver Stone. Amidst the stories of Snowden's escape and NSA wiretapping, one of the revelations to receive less attention at the time revolved around the use of metadata in the US military's drone strike program.

Buried in Snowden's release was the description of a program that used phone metadata, unconfirmed by human intelligence, to sanction lethal drone strikes.[27] In short, this means that lethal strikes (assassinations) by drones are authorized based on the data produced by and captured from mobile phones and other devices with no human confirmation of the identity of the target. Thus, if you are the target of a metadata-determined strike, loaning your phone to your grandmother means that she will become the recipient of a strike. This account was confirmed by a separate whistleblower, an operator at Joint Special Operations Command, whose account was released in *The Intercept*. That whistleblower described this practice as one of the primary causes of civilian deaths, noting that sometimes "it isn't until several months or years later that you all of a sudden realize that the entire time you thought you were going after this really hot target, … [and] it was his mother's phone the whole time" (Scahill 2015).

What these techno-socially produced moments have in common is a (re-)emergence of conscious consideration of an otherwise taken-for-granted function of technology. Each occurred when data escaped their original intended purpose and became representative of the individual in some unexpected and potentially disastrous way. The revelation of both of these initiatives produced outrage at the time, and yet led to no substantial policy or cultural change. Girls Around Me was promptly pulled from the iTunes store. Foursquare changed the function of its API to foreclose the possibility of similar applications in the future, but went on to collect more data than before.[28] While the drone program has continued, it has also been the subject of continued protests both in the US and around the world.

Sadly, both cases also had precedents that grabbed headlines but similarly failed to spark substantial change. Two years before Girls Around Me, Please Rob Me satirically illustrated the utility of spatial check-in data for potential burglars. Local news reports were scandalized, but users kept checking in. Seven years before Edward Snowden emerged as a charismatic and enigmatic heroic lead for news reporting, *The New York Times* used NSA sources to cover the Bush administration's use of warrantless wire tappings and data mining of US citizens (Risen and Lichtblau 2005). And yet, a few years later, when Strava unexpectedly revealed the location of military bases, the popular press was shocked yet again.

It's not even as if Strava was the first exercise application mapped on a large scale. In 2014, Nathan Yao had successfully scraped and visualized the running routes of RunKeeper, producing arresting visualizations through relatively straightforward code (see Figures 1.5 and 1.6) and what he called "the tip of a very interesting iceberg" (Yao 2014).

By now, we've automated the process of checking in. In 2019, the Foursquare app re-emerged as a "city guide" that tracks the places you go in order to suggest new venues you might enjoy. In 2020, Foursquare attempted to repurpose that data to track the spread of COVID-19 (Foursquare 2020). When, of course, the inevitable press coverage of to whom this application is selling user data or the biases of its algorithmically tailored urban experiences emerge, we will no doubt see more shocked press accounts. And yet people will keep signing into Foursquare, Strava, Facebook, Google Maps, Snapchat and all the others.

Or perhaps not. Perhaps now is the moment to end the perpetual "who could have known"-ism and instead recognize the likely outcomes of

Figure 1.5 Inspired by Nikita Barsukov's work, Nathan Yao built these maps using public RunKeeper data. https://flowingdata.com/2014/02/05/where-people-run/ (last accessed July 2021). (Used with author's permission)

```
1   (1..194).each do |page|
2     url = "https://runkeeper.com/search/routes/#{page}?distance=
3     &location=tacoma&lon=-122.4&activityType=RUN&lat=47.25"
4
5     puts "#{page}. #{url}".green
6     paths.concat ResultsScrape.scrape(url)
7
8     sleep [1.1,2.2,3.3].sample
9   end
10
11  routes = RoutesScrape.scrape(paths)
```

Figure 1.6 Sample code to scrape RunKeeper's public routes for the city of Tacoma. Written in Ruby by Josh Gray working with Jim Thatcher, it demonstrates the ease with which data may be acquired for those with specific technical knowledge. (See https://github.com/DataResistance for more examples of our work; also, note this script may violate RunKeeper's terms of service)

new forms of spatial data analysis, extraction, and visualization within longer histories of our relations with technology under capitalism (Thatcher 2018).

In this chapter, we traced pathways by which technologies, and through them, data, have come to represent us, to speak for us. Beginning from the neoliberal myth of individualized technological empowerment, we traced lines of critical thought on technology up through the present moment. There, we found how data, algorithms, visualization, and analysis have come to represent us, to speak for us, at the individual level over and over again.

In the next chapter, we turn from the individual everyday experience to more closely examine how we as a society have succumbed to almost mythological narratives of technological development and solutions. These narratives, which frequently star the paradigmatic "leaders" of Silicon Valley, have produced a society in which we bear witness *of* our lives *to* our technologies, accepting through faith that new technosocial relations are superior.

2
What Are Our Data, and What Are They Worth?

The Petabyte age is different because more is different.

(Anderson 2008)

In June of 2008, then editor-in-chief of *WIRED* Chris Anderson wrote his article "The End of Theory: The Data Deluge Makes the Scientific Method Obsolete." Provocative on its face, the article remains a touchstone for understanding the hubris of "data revolution," having been cited over 2,000 times since publication. In scholarly literature on data, it serves as a shibboleth and foil by which authors signal their critical intents (Kitchin 2014; Thatcher et al. 2018). The durability and influence of this article speaks to its capture of a certain modernist, triumphalist zeitgeist that frames more data as inevitably and irrevocably better. Perhaps most (in)famously, Anderson claimed: "With enough data, the numbers speak for themselves." He argued that it was no longer necessary, or even interesting, to know *why* something occurred, only that it would or would not occur. In this chapter, we argue that such an orientation is emblematic of our current condition, a context in which data are some of the most valuable commodities in the world precisely because of a near theological faith in data's perceived ability to produce knowledge and, therein, the world. We do so first by situating the narratives that Silicon Valley firms tell about themselves within this coproduction of a data-world. Then we demonstrate that through this context, our very relations to technology have shifted from *speaking with* data to situations in which we bear witness of ourselves such that data are made to *speak for* us:

A NEW commodity spawns a lucrative, fast-growing industry, prompting antitrust regulators to step in to restrain those who control its flow. A century ago, the resource in question was oil. Now similar concerns are being raised by the giants that deal in data, the oil of the digital era. These titans—Alphabet (Google's parent company), Amazon, Apple,

Facebook and Microsoft—look unstoppable. They are the five most valuable listed firms in the world. Their profits are surging ....

(*The Economist* 2017)

Setting aside attempts at regulation for the moment, we'd like to focus on the framing of data as commodity, and specifically as a *new* commodity. This framing is realized in particular ways for both tech giants and individual members of society. The simple metaphor "data is the new oil" has appeared in an array of popular, mainstream press stories about tech firms from *The Economist* above to *The New York Times* to *WIRED* (Dance et al. 2018; Matsakis 2019b). While there is a near universal consensus that data are *valuable*, such petro-nouveau claims are often unaccompanied by what exactly it is *about* data that makes them so valorized.

A thought experiment described by Antonio García Martínez, formerly of Facebook's monetization team, explains a bit of the conundrum behind data's valuation. In the pages of *WIRED*, he invites us to consider the difference between inheriting a tanker ship of crude oil and a van filled with hard drives containing all of Amazon's sales and browsing data for a year. The difference, he suggests, is that while he could sell the oil eventually, he would not be able to do the same with the sales and browsing data. The problem is that "Amazon's purchase data is worth an immense fortune ... *to Amazon*" (Martínez 2019).

This suggestion that such a data set could not be sold is, of course, bollocks, as Martínez goes on to admit later in the piece. Walmart, for example, would gladly purchase those hard drives, as would a number of other online and retail competitors to Amazon. That said, Martínez's overarching point that *data do not function the same as oil* is a sound one. What matters for us here is the mindset that brings him to that point. In his desire to explain why companies such as Facebook shouldn't have to pay individuals for their data, he misses the forest for his corporate benefactors' trees.

He's right that the revenue such technology companies generate is typically quite small *per user*. He states $25 per annum for Facebook globally and $130 for users within the United States. Other estimates suggest that on an individual level, data per user are worth much less that, perhaps only $1 annually (Steele 2020; Steel et al. 2013).[1] The huge revenues of technology companies come from their huge numbers of users, which is possible through the economies of scale that digital data and technologies make feasible. His revenue per user estimate misses the asymmetric relationship between Facebook and people. He pointedly ignores the extent to which the company sets the terms for how an individual's data are extracted even if the person is not a consenting Facebook user. This view of the world and people in it as a standing reserve of data to be monetized by the few who have the means to do so is dangerous.

Mark Andrejevic, a professor of media studies, describes this as the "big data divide." While the data are extracted from many individuals, they only emerge as having meaning and value for those who are able to collect, store, and analyze large volumes of such data (Andrejevic 2014). While you or I may be able to sell Amazon's van of hard drives, it's very unlikely that we would be able to put the data on those hard drives to much use without both access to advanced technical and computational infrastructures as well as some kind of business plan related to the buying and selling of goods online. It matters where, when, and for whom data emerge as commodities. Illustrating this point, geographer Jeremy Crampton and his co-authors have suggested that if the production of data can be seen as the production of value, then the object of our study should be not so much the content of the data, but the moments where "subjects are constituted as laborers in an exploitative economic system" of data production (Crampton et al. 2014, 3). Sociologist Christan Fuchs comes to a similar conclusion through a meticulous study of how social media users produce value for major technology firms under coercive, exploitative, unpaid circumstances (Fuchs 2014). These data economies work through processes of commodification and dispossession.

DATA COMMODIFICATION AND DISPOSSESSION

Scott Prudham (2009) suggests that something becomes a commodity when the purpose of its production is not an intended use, but instead market exchange (sales). Hand-drawing a map of town with a child is a fun way to explore the area with them. Hand-drawing a map of town for sale to tourists is producing a commodity. While both involve the same geographic landmarks, this distinction demonstrates the gap between daily practices that produce the spatial data points for direct use versus the production of spatial data for sale as commodities. A tourist map may sell for a few dollars; but when the data is global in scale, it may be valued in billions of dollars for those large corporations that can make use of it.[2] This gap between everyday use and commodity serves a function for tech companies as it hides the extractive nature of our relations with digital, spatial technologies:

> The individual [data point] produced from a single user at a given time and place (e.g., posting a picture of a meal to Instagram) is both nearly meaningless (Wilson, 2015[b]) and valueless (Stalder, 2014) until it is linked to the user's past data produced, the user's network of other users, the user's growing network of location data, and the temporal rhythms and spatial patterns embedded in data from many users. Conversion from an individual [data point] to an aggregated, digital commodity necessitates linking data across users, spaces, and times. These amalgamated data become necessarily large ("big") and thus a site for algorithmic selection, interpretation, and analysis as to what data to include and exclude.
>
> (Thatcher et al. 2016, 995)

This is a capital process in which companies colonize people's everyday lives through data.[3] Previously private times, spaces, and activities become subject to data (and specifically spatial data) extraction, analysis, and mediation by large companies for the sake of their business, with or without consent.

Following David Harvey's (2004) work on accumulation by dispossession, we use "data colonialism" as a metaphor that emphasizes the underlying processes of dispossession of data from those who create them. Further, we find the term useful as a means of highlighting the "wild west" ethos that continues to permeate much corporate action with respect to data privacy, access, and rights. However, it is important to note that colonialism was and is a horrific process by which peoples' lands, identities, and lives are stripped from them through systemic exploitation and appropriation. In their influential work on the topic, Couldry and Mejias (2019a, 2019b) argue for a non-metaphorical understanding of data colonialism. In short, they suggest that data colonialism calls attention to both the variegated ways that these processes play out across the globe and the ongoing ways in which colonial legacies support and undergird practices of modern capitalism. Noting the important debates around this definition (see, inter alia, Calzati 2020; Milan and Treré 2019; Halkort 2019; Dalton et al. 2016), in this book we focus upon the processes of dispossession that occur within data regimes. As such, we use the term "data capitalism" to refer to the overarching system in which acts of data dispossession (and resistance) occur. We metaphorically employ the term "data colonialism" to better call out exactly how data dispossession occurs and the ideological landscape in which data capitalism is forged.

Data colonization of everyday life is powerfully shaping our lifeworlds, which has direct implications for the choices we can make and the conditions in which we live and die. These effects come in two forms: First, data colonialism feeds the self-justifying mythos of moving fast and breaking things by suggesting that algorithms and data represent a new "wild west" to conquer. This frontier mentality builds on previous dispossessions to promise better living, and coincidentally revenue for tech companies, through data. Second, processes of data capitalism produce the individual that capital can see, and then feed that profile back to us through what we, following Melissa Gregg, call the data spectacle. This data representation shifts our relations with technology (and data) from *us speaking with them* to a relation of *them speaking for us*.

A NEW WILD WEST

Early internet evangelist Howard Rheingold was emphatic in declaring the internet a new frontier, one of infinite space and unlimited freedom (Rheingold 1993; Hirschorn 2010). Much like sanguine expressions of Manifest Destiny, such framings imagined the internet as an empty space to be filled by pioneers. Of course, just as the west of what is now the United States was filled with indigenous peoples with existing lives and relations to each other and the land, the spaces the internet came to fill were not empty. This is true both from an infrastructural perspective and in terms of our living relations in the world.

In *Code and Clay, Data and Dirt*, Shannon Mattern (2017, vii) starkly illustrates how "[i]nfrastructure begets infrastructure." [4] New systems tend to build upon old ones rather than recreate entirely new systems, no matter how "disruptive" they may claim to be. For example, Union Pacific, best known for operating North America's first transcontinental railroad in the nineteenth century, has been laying fiber-optic cable along its rail lines' rights of way for over three decades. By 2014, 34,000 miles of fiber-optic cable lay along Union Pacific's railroad rights of way (Johnston 2014). And Union Pacific is not alone. The overall network of internet cables in the United States tends to follow historic rail routes. [5]

According to the Chinese Railroad Workers in North America Project at Stanford University, upwards of 1,200 Chinese laborers died connecting the first transcontinental rail line in the US, and that is to say nothing of the dispossession and massacre of native peoples that occurred in order to secure the land along which the line itself runs. We may now have the means to video chat with family thousands of miles away, but the networks that allow that are built upon the expropriated labor of migrant workers and run through the stolen land of indigenous peoples.

Capitalism must grow to survive, finding new ways and geographies to accumulate value. At times, that growth happens along literal geographic frontiers, producing new spaces of commodities, market exchange, and exploitation (Harvey 1999). Under other circumstances, capital grows inward, altering everyday practices and places, commodifying and re-casting them in newly marketized relationships. In 2012, Jonathan Beller wrote:

> Capital's geographical expansion outwards is accompanied by a corporeal corkscrewing inward. Therefore, the visual, the cultural, the imaginal and the digital—as the de/re-terriorialisation of plantation and factory dressage, Protestant ethics, manners and the like—are functionalised as gradients of control over production and necessarily therefore of struggle.
>
> (Beller 2012, 8)

And just as the material infrastructure on which the internet is built is often elided, so too are the sources of data. This happens even as data are procured and analyzed to produce the digital representations that have come to speak for us in modern technological systems. This is a society of control through images, of commodified spectacles that offer false choices and further imbricate themselves into our very bodies. For technology firms, our bodily movements are empty spaces to be colonized by slipping data-generating practices into our daily lives. Google Maps renders movement and navigation into location-based advertising even when no billboard is in sight. Nest does the same with home thermostats. Tinder allows for dating to become a site for data generation modulated by digital representations of our identity and place. Our previously private acts, the banal and personal moments of our everyday, become imbricated within systems of capital exchange through the creation, extraction, and analysis of the data.

One of the great promises of modern software applications is to take everyday tasks, these moments of everyday life, and transform them into something easier, more comfortable, and preferably automated. Digital assistants such as Amazon's Alexa, Apple's Siri, and Google's Assistant, aren't simply predicated on answering your questions, but reflect a stated desire to provide you with solutions *before* you ask for them. A trio of quotes from former Google CEO Eric Schmidt illustrates this point:

> With your permission, you give us more information about you, about your friends, and we can improve the quality of our searches. We don't need you to type at all. We know where you are. We know where you've been. We can more or less know what you're thinking about.
>
> (quoted in Thompson 2010)

> The technology will be so good it will be very hard for people to watch or consume something that has not in some sense been tailored for them.
>
> (quoted in Jenkins 2010)

> I actually think most people don't want Google to answer their questions. They want Google to tell them what they should be doing next.
>
> (quoted in Jenkins 2010)

While the expression of this ideal varies from firm to firm, its underlying ethos has two important aspects. First, it is an invocation of the high priests of data, the idea that "data will make you better, because you are data." Moreover, this belief is realized in our daily rituals of truth-seeking, from searching Google Maps for a local store to checking what people are tweeting today (Hillis et al. 2013). In thought and practice, this reflects a fundamental shift from speaking with our data to them speaking for us. Second, it highlights a world in which, according to Ruha Benjamin, an attitude that seeks to "disrupt" life and convert it from "analog to digital" (Benjamin 2019b, 13) pervades the worlds' largest corporations and governments, but does so without consideration of "*the people and places broken in the process*" (Benjamin 2019b, 15; original emphasis).

BEARING WITNESS TO THE DATA SPECTACLE

Under data capitalism, people are represented by concoctions of captured data, the individual that capital can see. The data produced through daily life, through social media posts, credit card transactions, GPS tracking, and the like become the data that represent us within systems of capital flow and exchange. But agents of capital are not the only subjects in play. These representations (of ourselves) are also fed back to us and others in a spectacular fashion, distributed in a myriad of ways through everyday life.

Before turning to the specifics of the data spectacle, it is necessary to briefly introduce the concept of the spectacle as derived from the works of the Situationist International and one of their key thinkers, Guy Debord, whom we introduced in Chapter 1. The works of the Situationists are of particular importance for, as McKenzie Wark argues, it was historically "Henri Lefebvre and the Situationists that moved the site of Marxist critique from the factory to everyday life," requiring in turn a refocusing of critique upon political economies' "quotidian articles of faith" (Galloway et al. 2014, 195).

Perhaps the best-known idea developed by the Situationist International was that of the spectacle, the way capitalism separates us from actual lived experience and replaces those experiences with commodified representations thereof. Under these circumstances, the commodity form is "no longer something that enters into the sphere of experience in fulfilling particular needs or desires, but has itself become the constituent of the world of experience" (Chu and Sanyal 2015, 399). Building on Walter Benjamin's work on living spaces, spectacle is "the generalization of private life" (Lefebvre, in Wark 2011, 104). Commodities and associated market exchange no longer simply disrupt daily lives, they are the means through which the world is experienced. But this spectacular vision is not the full or complete world: "Apprehended in a *partial* way, reality unfolds in a new generality as a [commodified] pseudo-world apart, solely as an object of contemplation" (Debord 1967, 12; original emphasis). By replacing lived lives with commodities, what is *possible* to imagine, see, and do becomes delimited to what is *permitted* by market exchange (Debord 1967, 13).

In Situationist writing, spectacle is often articulated in terms of visual art and media, but the concept of the spectacle is meant to be much broader. In our current context, it provides a handy conceptual tool for better understanding the roles of technology and data in current societies. Melissa Gregg, Chief Technologist for User Experience and Sustainability at Intel, connects the concept of the spectacle to data. She problematizes performative representations of data as visual spectacle, such as in pitches for "big data" services. This gives rise to unsuitable ocular metaphors, such as data shadows, that distance a seemingly empowered agent from their data (Gregg 2015). Following Debord, we extend her critical analysis to incorporate the role of data as commodities, constituting a data spectacle that extends beyond the visual and that is intimately connected to the information political economy of our times (Thatcher and Dalton 2017).

With the advent of "big data," and specifically of spatial data of the everyday, a key article of faith in tech industries is the very representation and reproduction of self within those data systems cum spectacles. Data are valuable for multiple business models, so there is incentive to colonize everyday life to extract people's data, particularly spatial data. Tech companies use these data to accumulate value chiefly through targeted advertising, but also consumer and business services such as insurance and credit ratings.

The data spectacle comes into play when data are combined, processed, analyzed, and fed back to those from whom it was extracted and their peers. For example, instead of walking around a neighborhood to learn about an area, Craig tends to use apps on his phone to assist and augment his practice. The actions of his daily life have been colonized through the extraction of data by that device. Based on his location and multiple other data about him, those location-based apps present him a spectacle of his surroundings, algorithmically selected local features, advertisements, and reviews. These are likely relevant to him, but they also serve the app owner's business plan. These suggestions, inputs, and images are more than possibilities he chooses from freely. They shape how he sees the area and what he perceives as options, thereby effectively delimiting his actions to what the app's owners deem profitable to present. This is the data spectacle in action. Perhaps there are other things to see and do, and perhaps he may end up outside of the spectacle, but it is hard to know they are there, and much easier to not look. The data spectacle presents a commodified, fun-house mirror of the world and ourselves, exaggerating some things, minimizing others, and not showing things outside the frame. But the realities of the data spectacle are far from fun for many people.

These regimes of data have consequences for people on the ground as they see targeted ads and are subject to digital redlining (Noble 2018). Data-facilitated algorithmic analyses determine what sorts of coupons are offered and which businesses in which neighborhoods are recommended—or not. It takes the form of everything from targeted advertising to loan interest rates, product scores to projected recidivism scores. Horror stories of employers checking job applicants' social media histories have emerged as late-modern cautionary tales, as have a litany of guides on how to manage your social media presence to better land a job. More insidiously, social media have been suggested as an invaluable tool in determining creditworthiness. In a troubling article titled "Credit Scoring with Social Network Data," Wei et al. (2016) suggest social media data can be an invaluable tool in determining credit worthiness. In an interview on the matter, the authors lament that it is regulation to prevent discrimination that prevents the full embrace of such approaches within the United States (Knowledge@Wharton 2014).

This isn't to suggest that Wei and his colleagues *want* discriminatory systems. From their perspective, the use of social media data enables quite the opposite. Because more data make it so much easier to "run the numbers and figure out whether you're a good person," it would potentially, in their view, open up credit to swathes of the world currently locked out of traditional opportunities (Knowledge@Wharton 2014). Perhaps so, but while on one level it returns us to Ruha Benjamin's astute observation of the people and places broken by disruptive new technological practices, on another it rests firmly upon faith. In this case, the belief that the data you produce is an accurate and valid stand-in for you: that it can systematically indicate whether you are a good person or not.

Extracting our data and serving them back in the form of various commodified options facilitates crediting or blaming ourselves (not the system) for the consequences we experience and feel. Under the data spectacle, what can we do but perform better, in those data-fied terms, in the future? How better to measure that improvement than with more data, more quantification? Humans, if nothing else, are exceptional at manipulating patterns, at both conforming to and exploiting rule systems. We are so good at this that we can become susceptible to apophenia, seeing and attempting to manipulate patterns and connections between things that don't exist. Ultimately, this is what occurs as we attempt to perform ourselves in the fun-house mirror of the data spectacle by using the various technological apparatuses with which we interface daily. Melissa Gregg (2015) uses the delightfully visceral term "data sweat" for data that are essentially of us, but may exude in ways we don't quite intend. She goes on to suggest that:

> labor we engage in as we exercise and exchange our data—especially in our efforts to clean up our image, present a hygienic picture, and make ourselves look good—is a kind of sweat equity for the digital economy. It is a form of work we perform in the attempt to control what is ultimately out of our capacity.
>
> (Gregg 2015, 45).

While we may try to alter or curate those spectacularized, data-fied (re)presentations of ourselves to better control our dating, financial, professional, gastronomical, and other experiences, we are struggling against an ultimately unknowable force. The proprietary algorithms which sort and shape us, whether via the locations we may visit, recorded in Google Maps and Facebook profiles, or what we've posted on Twitter, are ultimately trade secrets.

We have a shared vested stake in the (re)presentations of ourselves by which corporations, other people, and ourselves come to know us, but currently the means are shrouded in mystery. Our actions to manipulate hiring algorithms are more than apophenia, as an algorithmic pattern does exist, it's just (relatively) unknowable to us. For the high priests of data within major firms, this doesn't matter. After all, as Chris Anderson wrote, it doesn't matter why something happens, just that it does. Amazon is not so interested in why you might buy a purple cashmere sweater as it knows when and for how much you will.

Troublingly, this type of reduction is also a key tenet of how "smart cities" function and are governed, according to sociologist Jennifer Gabrys. Drawing from the work of Michel Foucault and with a bit of admitted irony, she coins the term "biopolitics 2.0" (Gabrys 2016, 192). While for Foucault biopolitics was about control/governance over the milieu in which human beings lived, of their ways and practices of life (Foucault 2003), Gabrys adroitly inserts the "2.0" to emphasize the milieu of humanity, technology, and environment within smart cities and smart developments. If our future is to be "smart," she highlights the dangerous "transformation of citizens to data-gathering nodes" and how it "potentially focuses the complexity of civic action toward a relatively reductive if legible set of actions" (Gabrys 2016, 203). Efficiency, predictability, and increased production become the solutions to the challenges faced by urban residents and local governments.

SOCIAL CREDIT

While smart cities move in this direction, initiatives like China's social credit system drive "biopolitics 2.0" to its logical conclusion. The intent with this kind of system is to algorithmically calculate a reputation or "trustworthiness" score for each adult citizen based on multiple indicators which may include taxes, debts, crimes or citations, purchasing histories, and community service. Not paying debts, crimes, minor rule-breaking such as jaywalking or eating on public transit, and even too much time playing video games could decrease the score. Actions deemed as positive for society, such as caring for the elderly, donating blood, community service, and raising a child would increase it. Too low a score results in punitive measures, which may include being banned from commercial air travel and high-speed trains, being prohibited from receiving a loan or purchasing property, exclusion of one's children from admission to desirable schools, and even potentially public shaming, such as seeing one's name and face on billboards listing "untrustworthy" individuals (Vanek Smith and Garcia 2018). Those with a high score enjoy benefits which may include shorter waiting times for government services, easier access to credit, and even the option to publicize their score on dating services (Kobie 2019). While social credit could in the future become a unified system at the national scale, Chinese social credit currently exists as a web of multiple different systems. Some are thematically focused, including the justice system, while others are regional, working only in particular cities or provinces. Still others, such as Alibaba's Sesame Credit, run through contracts with major private companies.

Regardless of whether the ultimate intent is direct or merely symbolic control, the extent of such a system is only possible by collecting data from multiple sources and calculating them in an automatic fashion.[6] Once again, the resulting numbers are assumed to speak for themselves. The question of "why," much less the contributing social circumstances for a score, is not deemed important. Unsurprisingly, such systems are ripe for abuse by unscrupulous authorities, unintentional bureaucratic errors, or cultural biases, any of which can have dramatic consequences for someone's life.

This all sounds dystopian, and it is. That point is belabored by many breathless English-language press accounts (Matsakis 2019a; Mozur 2018). Nevertheless, it bears mention that similar social mechanisms employing algorithmic calculation of personal data are already in place in the United States, Europe, and elsewhere. Our debts are monitored not only by banks, but also by entirely unaccountable credit rating companies such as Experian and TransUnion. Insurance companies assess risk by monitoring driving and attempt to incentivize and monitor exercise and other healthy behavior (see, for example, Progressive's "Snapshot" discount for "good" drivers). Criminal justice, from policing focusing on minor infractions to sentencing to incarceration, and even public shaming on the internet, is increasingly algorithmically guided (Eubanks 2018). Travel is limited not by official prohibition, but through the market.[7] Similarly, money provides access to services and legal counsel, largely preventing those with means from having to deal with waiting for government bureaucracies. The implications of falling down due to initially minor mistakes, uncontrollable circumstances, incorrect data, or biased algorithms are no less extreme in North America or Europe. What makes the Chinese system different is its use in the context of a non-democratic government.

LAW ENFORCEMENT SURVEILLANCE

Whether in China, the United States, Europe, or elsewhere, when social credit's push comes to shove, law enforcement tends to get involved. Though often formally enacted by government agencies, here again data dispossession and its expansionist capital imperatives enter the picture. For example, due to recent US Supreme Court cases, American law enforcement agents must get an appropriate warrant before collecting a suspect's GPS location history or requesting it from a mobile phone network provider.[8] Yet today, law enforcement and national security agents are increasingly circumventing these rules by simply purchasing the very same data instead. Similar data markets exist for license plate scanner data and facial recognition, circumventing attempts at regulating the use of said algorithms and data (Cox 2021). Moreover, in an age of social media, some revealing data are simply available on social media platforms. For example, as Black Lives Matter protests swept the United States in the wake of George Floyd's murder at the hands of police officers in 2020, police made clear their ability to track down protesters after the fact using digital data. For example, when two police cars in Philadelphia were set on fire:

> FBI agents were able to identify [the alleged perpetrator] thanks to an investigation that largely relied on data freely available online, based on an aerial video taken the day of the protests, an Instagram picture, photos taken by an amateur photographer, and—crucially—a forearm tattoo and an Etsy t-shirt.
>
> (Franceschi-Bicchierai 2020)

According to press releases, this particular investigation was done through human detective work, even as it relied on digital data available online. Similar methods were used to identify right-wing rioters who stormed the US Capitol building in January 2021. These examples stand out because law enforcement press releases in the United States often elide the use of supporting digital surveillance, such as the use of stingray devices to identify all the mobile phones in a particular area at a particular time. Furthermore, law enforcement's use of algorithmic services to identify suspects has grown astronomically in recent years, even as false positives and racial biases in the services have emerged (Hill 2020). Examples include Clearview AI's facial recognition service that draws on a massive database of photographs scraped from the web and associated social media.

THE NEW NORMAL

Given these developments in the United States, China's burgeoning surveillance networks serve as an orientalist "black mirror" for western media outlets. Paul Mozur's 2018 piece in *The New York Times*, "Inside China's Dystopian Dreams: A.I., Shame and Lots of Cameras," speaks of "a high-tech authoritarian future" in which "China is reversing the commonly held vision of technology as a great democratizer." Mozur's account obfuscates any reflective consideration of conditions in the west, for example pointing out that China has roughly four times as many surveillance cameras as the United States, but leaving out that it also has more than four times its population. As of 2019, the United States had the most surveillance cameras per person in the world, according to PreciseSecurity (Baltrusaitis 2019). Similarly, the existence of a list of 20–30 million individuals suspected of criminal activity in China is high-

Figure 2.1   One of Chicago's police surveillance cameras. Through "Operation Virtual Shield," the city has access to more than 10,000 surveillance cameras, many with artificial intelligence-enhanced facial recognition and automatic tracking of individuals and vehicles. The ACLU of Illinois (2011, 2) notes that these abilities "far exceed the powers of ordinary human observation and dramatically increase the power of the government to watch the public." ("Chicago Police Camera," licensed under CC BY 2.0. https:// creativecommons.org/licenses/by/2.0/#, last accessed July 2021)

lighted, but little mention is given to the Interpol Terrorist Watch List, the US Department of Homeland Security's No Fly List, or any of the other myriad lists that have tracked millions of individuals for decades in other countries. *No, this is new and different.* The fact that it has already and continues to occur within other nations is immaterial, that it is in line with data extraction and analysis efforts by state and private industries around the globe is ignored; here, China remains the other.

And then, in late 2019, COVID-19 emerged in Wuhan. As our societies and lives continue to writhe and contort as global capitalism attempts to survive a crisis it not only manufactured, but continues to perpetuate, something interesting happened in China. Surveillance systems were repurposed to track the disease, and a new "black mirror" emerged. Systems previously used for general surveillance were repurposed for the ongoing epidemic. Contract tracing through public streets and identifying likely symptoms on corners became tasks of systems designed for more general purposes.

And yet, even as outlets like *The Guardian* suggest (likely correctly) that China's increased surveillance to track COVID-19 may likely become "the new normal," left out are the similarities (and failures) of similar systems in other nations.[9] While both individuals and algorithms pored over camera footage in China, around the globe another source of data became intrinsic to supposed disease response: as always, mobile phones. From Norway to the United States to, of course, China, mobile phone location data were touted as a means of tracking and eliminating the disease.

It didn't work. Privacy concerns shut down the mobile app in Norway, while in the United States the tracking never materialized. Mobile phones stand in for us in so many ways, up to and including as authorization for drone strikes by the United States government, but there remains a gap between what they can capture and our emplaced, visceral bodies. The spatial resolution of location tracking exists at a scale other than that necessary to control an airborne virus.

SPEAKING FOR

How, then, can we make geographical, technological systems work less as bureaucratic fixes and more for actual people? We suggest that it is the very opacity of algorithms and the spectacular (re)presentation of self that have transformed the relationship from *speaking with* to *speaking for*. In a provocative piece on the need for "big theory" to address big data, anthropologist Tom Boellstorff (2013) writes:

> The confession is a modern mode of making data, an incitement to discourse we might now term an incitement to disclose. It is profoundly dialogical: one confesses to a powerful Other. This can be technologically mediated ….

As we've discussed and demonstrated, on one level, quantification and the production of data have come to colonize our life-worlds with the intent to extract value. On another, they do so in exchange for "notional advantages, [spectacular] experiences in which aspects of [our] lives are algorithmically sorted and produced for [us]" (Thatcher et al. 2016, 999). However, these advantages are offered in an opaque manner. The algorithms are trade secrets, and any ability to manipulate them occurs at the risk of our tendencies toward apophenia.

The asymmetric relations between the algorithms which increasingly guide our lives and the actions we take are mediated through the data we produce. That key moment of quantification also serves as the moment in which our relations with ourselves become one of faith. Much like the supplicant whose prayers are answered after a sacrifice, we keep doing those things which *seem* to produce results for our (digital) selves.

For example, a local politician in the midst of a campaign several years back revealed to Jim how he only put up social media posts during two two-hour windows: at the beginning and end of the workday. He believed this was a time when people were browsing and therefore more likely to see, like, and share his posts. Was this true? Maybe. A whole cottage industry exists to help us with our digital presentations. A firm called Later, dedicated to Instagram marketing, claims that the ideal time to post is 9–11 a.m. Eastern Standard Time (with caveats for local conditions and businesses). They based this on their analysis of "12 million Instagram posts, posted in multiple time zones around the world from accounts ranging from 100 to 1 million+ followers" (Chacon 2021). However, Later also emphasized that "it's best to find your *personalized* best times to post," in essence admitting that while big data may tell us one thing, specificity—the *why*—remains unknowable. For Anderson and his ilk, this doesn't matter. For advertisers with global reach, the specifics matter less than the outcome. But for us? For individuals who will be policed, get dates, find restaurants, and be offered jobs based on the data we produce?

For us, those things matter quite a lot, and so we're left with our propensity for apophenia in the face of the data spectacle. Our data sweat, the labor we undertake to produce and control our digital selves, is more about guesswork than control. We've become supplicants at an altar to unknowable data gods. We post what we think will look good, we perform our humanity in the face of an apparatus over which we have no control and little understanding. In this way, we *bear witness* to the selves that our digital data produce, rather than having any concrete control over their (re)presentation and (re)production. Our machines *speak for* us rather than *speaking with* us. When that speech aligns with our prayers or hopes, a match on Tinder, a job offer, a mortgage, we reproduce those actions, hoping that the past will predict the future—praying that we've found a confession that works.

This is what we, along with others, suggest forms part of a spectacular "howling feedback loop" in which "[t]he data generated by such actions is then fed into systems which algorithmically shape what options will be presented the next time (s)he makes use of the service" (Thatcher and Dalton 2017, 140; Lohr 2012; Wilson 2012).

This may feel benign or seem (or even be) beneficial in certain circumstances. Individuals may have preferences for specific kinds of food. If a food-delivery service notices that Craig tends to get pizza not nearby, but at Mama Rosa's in the next neighborhood over, it can be helpful for him to get coupons or promotions for Mama Rosa's rather than pizzerias closer to his home. With modern machine-learning, this occurs without the algorithm ever needing (or trying) to understand *why*, it simply detects a pattern along some dimensional axes of data (Mackenzie 2017). As long as coupons are the upshot, it does not matter if it is because Mama Rosa's pizza is better or because Mama Rosa's happens to be along his commute home.

UNEVEN DATA

And yet what opportunities (and peoples) are foreclosed from consideration by these loops? Information Studies scholar Safiya Umoja Noble argues that while it is "certainly laudable" to suggest that major technology firms like "Google/Alphabet [have] the potential to be democraticizing force[s]," it is necessary to consider both who benefits from data and algorithmic practices as well as how the effects of their adoption are not experienced equally across populations (Noble 2018, 163–164). Someone without access to credit may not be able to use a credit card to access pizza deliveries. Not only can some people fall through the categorical and algorithmic cracks, social biases are replicated and catalyzed by such systems. In *Automating Inequality*, one of the many individuals Virginia Eubanks spoke to echoes this sentiment, suggesting that more privileged individuals should pay attention to the surveillance and algorithmic governance enforced upon the less privileged because they, the privileged, will be next (Eubanks 2018). Ruha Benjamin puts it more bluntly when she writes that "Black people *already* live in the future" (Benjamin 2019b, 32; original emphasis).

Whether we focus our attention on race, gender identity, age, income, or geographical location, who is counted and how they count in data plays out unevenly. Writing several years ago with Linnet Taylor, we discussed the "uneven development of data," how the data profile of a mobile device user in Mauritania would likely differ significantly from that of someone in central London (Dalton et al. 2016, 1). As we have seen, this difference not only affects what these individuals can come to know within digital systems (search results are tailored by location and profile), but also their very access to opportunities like investment capital.

This is why the "decision" to participate within these systems, to exchange our data for their notional advantages, is such a disingenuous framing. On the one hand, there are real material advantages to the acts of participation that play out differently across different populations and places. Ordering groceries for delivery is very helpful during a pandemic. On the other hand, the very *choice* is one only offered to those who have access to the advantages through other means. Keanu Reeves can famously use a flip phone because he has a coterie of assistants that can handle his financial and personal needs. A single mother who has just been laid off may be forced into accepting a variety of surveillance and data generation processes simply to access financial support for her family.

The ability to opt out, to the degree that it does exist, favors those whose lives are not predicated upon participation. Inclusion and exclusion within data generating systems, and the algorithmic results they produce, matter. The stakes of actual resistance, as opposed to privileged side-stepping, are steep. They may result in denied health benefits, inability to access needed financial resources, or even refusal to be considered for employment or housing.

In the next two chapters, we survey practices of resistance and put them into conversation with the larger project of solidarity through machines—a shift back from *speaking for* to *speaking with*.

3
Existing Everyday Resistances

Where there is power, there is resistance.

(Foucault 1990, 95)

In early 2020, artist Simon Weckert became a traffic jam all by himself. He did not cause a traffic jam by blocking a street; rather, he appeared as a traffic jam on Google Maps, causing cars on the surrounding blocks to be routed around the street he was on. Weckert's jam was a simple art performance/installation he called "Google Maps Hacks."

> 99 second hand smartphones are transported in a handcart to generate [a] virtual traffic jam in Google Maps. Through this activity, it is possible to turn a green street red, influencing the physical world by navigating cars on another route to avoid being stuck in traffic.
>
> (Weckert 2020)

Weckert's "hack" works by leveraging how the Google Maps app tracks individuals to generate its traffic information. If location services are enabled on a mobile device, especially when turn-by-turn navigation is active, the company is collecting location and speed information from that device. Those data are then fed into a system that estimates traffic conditions based on the aggregate movement of phones along that street. Weckert thought of his concept during a May Day rally in Berlin, when he noticed that his Google Maps application was assuming that the people in the streets were cars and therefore symbolizing them as a slow-moving traffic jam. As he notes, this intersection of material and virtual is a powerful forum for "performance of activism" (Goldstein 2020).

Weckert's actions are amusing and inspiring, but it doesn't take an artist or an activist for someone to exert greater control over the data they produce and how it is extracted from them. In this chapter, we survey different individualized ways of responding to the geographic data collection in everyday lives. Here, the focus is on tactics: small-scale, everyday actions that resist, contest, and alter the production and extraction of geographic data, such as Weckert's traffic jam.[1] We begin by exploring the shortcomings of privacy measures offered by technology companies to address concerns about current modes of data extraction and the too-common rhetorical binary choice of privacy or security. Such privacy-washing offers the impression of protection while maintaining the underlying system of data extraction and exploitation. From there, we explore several kinds of more independent reactions to data dispossession to develop a cohesive, but not necessarily comprehensive, typology: *acceptance*, *active resistance*, *making present*, and *escape*.

In practice, such choices and tactical actions are highly contextually dependent, often deeply personal, and produce decidedly mixed results. No one engages in all of them. Rather, most adopt a practice to a degree, and may mix it with others. As individualized everyday practices, none present total or complete engagements with the scope of data capitalism. Furthermore, the very ability to engage in these acts rests upon variegated layers of privilege and ability. Who someone is will have bearing on what tactics are possible for them. That these lines of resistance are not equally available reflects and contributes to broader processes of prejudice and dispossession within digital and material environments (Eubanks 2018; Benjamin 2019a). Some forms of escape, for example, are simply not an option for Black Americans in heavily surveilled neighborhoods or for residents of economically struggling rural areas with spotty cell phone and broadband coverage. Likewise, it is increasingly common for employers, especially those in the so-called "gig economy," to require workers to provide data about themselves; if only one provider will offer a contract, the pathways of resistance can be rapidly curtailed by the need for employment.

In addition to the inequities of access, individualized tactics are inherently limited in what they are likely to accomplish unless they are part of broader, more collective efforts. One person shutting down their Facebook account will not lead to a meaningful privacy policy. Employing GPS spoofing on a phone or a virtual private network (VPN) over an internet connection may provide some additional individualized privacy, but it does little to alter the social imperatives and strategies of data capitalism and the technologies they inspire. Even in the uncommon event that an individualized privacy method becomes popular enough to impact the market, the incentive is for companies to develop countermeasures to keep the data gravy train rolling. For example, while web browser ad-blocking software has become common, even built-in to some browsers, data firms also continue to find ways to collect data by other means (Nield 2021). Broad, meaningful changes in the design of data technologies and related policy require more coordinated efforts that engage not only the economies and politics of data, but also the cultures and norms around them.

Even with all the biases and structural limitations, everyday tactics can offer a degree of data control that is comparatively easy to implement. It probably won't save the world, but it might help you.

PRIVACY-WASHING IS NOT ENOUGH

Read at face value, technology firms' press releases make the case that their top priority is users' privacy and that it has never been easier for users to control how their data are collected and shared. Of course, the seemingly endless succession of data scandals (Bishop 2018), threats of regulations and fines (Information Commissioner's Office 2020), and antitrust and class action lawsuits (Georgiadis and Beioley 2021) suggest that other factors are at play in these promises.[2] Whatever the motivation, some firms are beginning to allow individual users some control and intentionality in how each user's data are collected and shared. Examples include app-specific access control to GPS information and the amount of time before location history data are deleted from a Google account (Morrison 2020).

However, not only does such language deceptively shift responsibility towards the individual, it also does nothing to alter the underlying business model and profit incentives on which these firms rest. Data capitalism is built upon the creation, extraction, appropriation, analysis, and trade of data. Privacy-washing provides the appearance of conscientious responsibility to ensure the continued profitability of this strategy. For example, Facebook, currently the world's largest social media company, has been most brazen in its privacy-washing attempts and, as such, has drawn a great deal of media scrutiny and proposed regulatory action. The company's Cambridge Analytica scandal and its media coverage illustrate both how responsibility is shifted to the individual and how the profit motive drives considerations of "privacy" within systems of data dispossession.

Cambridge Analytica was a third-party research and data broker firm that improperly obtained upwards of 87 million Facebook users' personal data which it then sold to predominantly (though not exclusively) conservative political campaigns and organizations including the 2016 US presidential campaigns of Ted Cruz and Donald Trump, and the pro-Brexit campaign in the UK. Early coverage in *The Guardian* highlighted how the data were obtained and sold "largely without [the users'] permission" (Davies 2015), but the story did not explode into a full-blown scandal until years later, when *The New York Times*, *The Observer* (London), and *The Guardian* received a trove of documents detailing Cambridge Analytica's practices from an internal whistleblower (Confessore 2018).

Once the scandal broke, Facebook shares plunged, losing approximately $119 billion in market value, the largest single-day loss by one firm in Wall Street history to date (Picchi 2018). Narrating the companies' loss of users and expected downturn in revenue, the company's chief financial officer, David Wehner, explained that giving users "more choices around data privacy" after the scandal was a likely culprit for the drop (Neate 2018). In a Facebook post (and echoed in a *New York Times* interview the same week), Facebook CEO Mark Zuckerberg similarly outlined the company's response and responsibility in three main points: (1) Facebook had already changed third-party access permissions to user data in 2014, but would conduct a full audit to establish whether other breaches had occurred; (2) third-party developers' access would be restricted even further; (3) it was up to users to fully understand what applications were doing with their data and to revoke permissions if they objected. In short, Facebook had already addressed the issue, but they were nobly willing to sacrifice their own profits to make sure it didn't happen again, and ultimately it's your responsibility anyway. The first, and as of this writing, only update to the internal audit came two months later, and revealed that "around 200" applications had already been suspended for being in violation of data privacy practices (Archibong 2018).

Several months later, Facebook was hit with another wave of privacy breaches as *The New York Times* revealed that it had failed to monitor how device manufacturers handled the data of hundreds of millions of users to which Facebook had granted access (Confessore et al. 2018). This time, despite coverage in *The New York Times* and pressure from US Senators, little came of the issue. The best explanation appears in Brian X. Chen's (2018) article, also in *The New York Times*, "How to Protect Yourself (and Your Friends) on Facebook." As he wrote:

What, if anything, can [users] do to protect their data connected to the social network?

Here's the hard truth: Not much, short of ceasing to browse the web entirely or deleting your Facebook account.

Ultimately, the asymmetric nature of the relationship between user and technology firm ensures that while "best practices" may be followed, there is little actual control within the system itself. Corporations will privacy-wash their practices, particularly when their revenues are at stake, offering more (mostly meaningless) choices and greater (largely fictional) control. Moreover, with approximately 2.45 billion monthly active Facebook users in January of 2020, the stakes for how people engage Facebook and similar services are clearly high. Technology companies cannot be relied upon to offer users greater control over their data. Thankfully, such companies and their services do not wholly determine the actions of their users. As we'll see, users may repurpose a service, and even resist the intentions of its designers.

EVERYDAY PRACTICES

In popular media, the dispossession of personal data is frequently framed as a binary tradeoff between security (through surveillance) and privacy. Moreover, the choice is often presented as resting with a technology's designers. Perhaps the most common, reductive form of this argument was stated by the CEO of Google at the time, Eric Schmidt, when he claimed: "If you've got nothing to hide, you've got nothing to fear." The totalizing binary is presented as a choice of one or the other: privacy (in this case, a nefarious one) or surveillance (in this case, a protective one). While much more nuanced discussions exist around privacy and security,[3] this diametric framing still structures much popular media discourse on the topic. For example, when the Norwegian Data Protection Authority temporarily banned the processing of data from that nation's COVID-19 tracking application, named Smittestopp, over concerns about the invasive nature of the data gathered ("privacy"), the Institute of Public Health argued that such measures weakened the ability to control the ongoing pandemic ("security") (Treloar 2020).

In actual lived practice, such distinctions are rarely so simple. Even in the most straightforward cases, the privacy/security framing runs afoul of two major assumptions: first, that a data technology actually works as advertised, and second, that consumers/users will only act in line with the designers' intentions. A closer examination of these assumptions reveals a dialectical tension, rather than a diametric choice, between "privacy" and "surveillance."

For the first, technology companies, and startups in particular, are notorious for promoting "vaporware," over-promising the functionalities and possibilities of their products to attract publicity and investors. Perhaps the most infamous example, Theranos, was once valued at $9 billion before collapsing under the weight of fraud and the fundamental impossibility of its various medical promises (Carreyrou 2018). Driverless cars may someday be commonplace, but as of this writing, the billions of dollars invested in their engineering and geographic data do not live up to the hype of technology company tycoons. For example, in July 2020, Tesla CEO Elon Musk anticipated a car that required no driver input by the end of that year (BBC News 2020).

Data technologies not only frequently fail to live up to popular expectations, they can, and often do hang, break, and crash. In our experience, GPS-enabled turn-by-turn navigation usually works as intended, but sometimes it just doesn't (Dalton and Thatcher 2019). Under the right circumstances, glitches can be opportunities for wildly different, even liberatory, experiences and modes of expression (Russell 2020). A technological navigational error could send us to a better (or worse) location with different possibilities than originally intended.

The second assumption, that users will always do as the designers intended, runs into problems when exploiting those glitches becomes intentional. Users will attempt to utilize a technology to fulfill their own needs and desires beyond what the technology's designers intended. However, users do not have total freedom, as their actions are limited by the material structure of technology developed by the designers for their own purposes: the users' margin of maneuver. A pool noodle can't hammer nails, no matter how hard you swing it. Users operate within a margin of maneuver between how they want to employ a piece of hardware or software and what is possible given its designed material structure. Under the right circumstances, within that margin of maneuver can be found the means to repurpose/refashion/redesign a technology that sparks subsequent larger technological and social changes. Using components from a junkyard, Grandmaster Flash (Joseph Sadler) developed an early cross-fader/DJ mixer, a vital tool in the early development of sampled music in general, and now an inexorable part of audio technology, Hip-hop art, and Black culture (Gaskins 2019). Similarly, self-driving cars provide potential opportunities for new kinds of repair and aftermarket automotive modifications (Alvarez León 2019).

Given the contexts and possibilities of practice, the absolutist tradeoff between surveillance and privacy breaks down. To get a better handle on our data, we need a more nuanced understanding of everyday technological practices, a dialectical one. There is no doubt that many users acquiesce to most technological conditions, but they do so in ways that are partial and personally articulated.

Here, we begin to categorize how individuals actively and passively resist and/or shape the collection and use of their geographic information. This typology is not intended to be as comprehensive as other less geographically focused work on resisting surveillance (Marx 2009). Nor is it as detailed and nuanced as the number of contexts and circumstances it begins to describe. The dual purpose in identifying these tactics is to begin approaching existing practices so that individuals may consider whether and how to apply them in their own contexts, and at another scale, begin to cobble together such practices into broader, more comprehensive moments of resistance and alternative politics. Indeed, we hope that this typology is not merely useful for a scholar seeking to understand peoples' actions, but to a popular audience as a way to learn the resistance methods that are available so they may be shared, refined, and used both individually and in groups.

Table 3.1   A typology of responses to data capitalism

Response	What is it?	Example
Acceptance	Consenting or acquiescing to data extraction, dispossession and analysis (business as usual)	Clicking "Accept" on the terms of service
Active Resistance	Attempts to assert some control over data extraction	Turn off location services, fabricated information/data poisoning, face coverings, VPNs, Tor, ad-blockers
Making Present	Attempts to make the mechanisms of data production, extraction, and analysis legible and understandable	Mapping infrastructure, contextualized exposés
Escape	Attempts to avoid generating or extracting data	Paying with cash, using a "dumb" phone

With those goals in mind, it is important to note that this is not a technical guide detailing specific methods for resisting data capital-ism. Such technical guides already exist (Thompson and Wezerek 2019; Goodin 2020). Moreover, these resources need to be frequently updated due to the rate of technological change, shifting conditions of govern-ment regulation, and the continuing arms race between personal privacy and data dispossession. We focus on broad *kinds* of actions, both because *specific* actions are highly contextual and in order to uncover new ways to think about and enact resistance and solidarity, to move from what is to what might be.

ACCEPTANCE

It is almost impossible to not acquiesce to the creation and extraction of data to some extent. Unsurprisingly, most of us do so: a 2018 Global Attitudes Survey conducted by Pew Research estimated that "more than 5 billion people have mobile devices, and over half of these con-nections are smartphones" (Silver 2019). In other words, over half of the global population are producing data that reveal their location and travel patterns to some degree.[4] But with an increasing number of social functions (dating, eating, travel, and so on) mediated through mobile applications, many, if not most, people simply check the "Agree" terms of service box and move on with their day, perhaps grumbling about it, but *what can one do?*

Clicking "Agree" is a crucial moment of acceptance that invokes an end-user license agreement (EULA), shaping what is legally permissi-ble by the company, even as the EULA itself is rarely read. In 2005, for example, the company PC Pitstop altered its software EULA to include a clause offering $1,000 to anyone who read the clause and contacted it. It took five months for someone to claim the prize (Magid 2009). Academic research, such as Lin et al. (2012), has also shown little actual engagement with EULAs before their acceptance. And yet they mark a pivotal moment in the creation, extraction, and control of the data we produce. As Thatcher et al. wrote in 2016 (996):

> Previously public—or, in this case, nonquantified—information about daily life is quantified and privatized, not in the hands of those who generated it, but of those who created the application; whether the espoused motivations for quantification are to enhance the

service or add value to the dataset being assembled, the transfer of ownership remains.

Before a user clicks "Yes" on an EULA, a corporation is limited in how it can collect data about people. In addition to any pertinent and enforced government regulations and legal liabilities, it faces practical, technical limits. It may not have access to a phone's contacts or the device's location history. However, after checking that box, the company is limited by the much broader terms set out in the agreement. EULAs are notorious for granting access to wide swathes of data collection that isn't directly related to the service rendered as well as granting corporate ownership of and the right to sell said data. In addition, many EULAs force users into arbitration for any disputes and grant the company the right to alter the terms of the agreement at any time. This asymmetric relation is so well known that the crass, satirical American television show *South Park* produced an entire episode mocking hidden terms in Apple's EULA, and yet users keep clicking "I agree." Why? There are two major, underlying, interrelated reasons.

First, EULAs are easy to ignore. It is easy to click the box and forget about it. The extent of geographic data production and expropriation is most visible not in the EULA, but in the media flash of a data scandal. Due to scandals, the public *know* that their data are being collected and sold, but the exact process and extent are often obfuscated in the EULA in ways that make it difficult to parse out. For example, Aleecia McDonald and Lorrie Cranor (2008, 563) estimated that "if all American Internet users were to annually read the online privacy policies word-for-word each time they visited a new site, the nation would spend about 54 billion hours reading privacy policies." That is approximately ten days per person per year.

This ignorance is purposeful and directly leveraged by technology companies. The agreements provide legal coverage for corporations to do what they want, buried in a morass of terms that require a law degree to fully understand and act on. Thus, for the rest of us, ignorance is used as a pretense of consent constituted by clicking "I agree." However, as more stories of data scandals in the news and calls for regulation emerge, the utility of ignorance may be falling. Users may not fully understand the legal terms and conditions, but such scandals build broad distrust of data-driven companies, even those that stay within their stated terms and conditions. Ignorance may have contributed to the growth of data-driven companies while they had untarnished reputations, leading to the data landscape we have today, but not knowing or understanding the contents of a EULA doesn't work if consumers broadly distrust the company offering the service.

Nevertheless, even in the current social landscape, the second, related reason still holds: we aren't really offered much of a choice, are we?

> The reason people click "yes" is not that they understand what they're doing, but that it is the only viable option other than boycotting a company in general, which is getting harder to do.
>
> (Lanier 2014, 314)

Increasingly broad, fundamental swathes of our existence are mediated through location-aware applications, from the romantic, such as Tinder and Grindr, to the economically mandatory (imagine handling your job without email). Tech companies offer an all or nothing choice. There is no negotiation with an EULA, we cannot agree to only certain parts or modify the terms. The benefits are immediate, such as the ability to purchase a commuter train ticket or take a photo and share it with friends, and the costs are hard to recognize. What does it mean to have my location history now owned by this technology firm? What will they do with it, when?

While there may be cracks in our trust of technology corporations, their ubiquity has (so far) prevented large-scale rejection of their practices. Particular moments, such as when the hashtag #deletefacebook trended on Twitter, are less a rejection of the terms set by technology firms and more about the shift between competing firms offering the same fundamental choice. In that case, it was leaving Facebook and announcing it on Twitter. This does not, however, mean that we cannot and do not resist.

ACTIVE RESISTANCE TO GEOGRAPHIC DATA COLLECTION

There are many reasons to acquiesce to the production and extraction of our spatial data, from the perceived tradeoff being worth it—"Sure, my location history is worth a free donut!" (Probrand 2019)—to the overwhelming asymmetric nature of the relation—"I need email to find a job, and the provider requires I accept these terms ...." While we may countenance individual instances as small, data accrues over time, and detailed profiles of our lives emerge, turning the balance in favor of the data broker. In contrast to the various ways of and levels at which we accept the expropriation of our emerging data selves, active resistance refers to attempts to directly impede the production or extraction of personal spatial data through greater engagement with the technology in question. Active resistance, in our typology, does not refer to efforts to avoid generating or extracting data (that is *escape*), but rather to a range of ever-shifting tactical practices that attempt to assert some control over those processes.

Much as nearly all of us accept data extraction to some degree, active resistance is similarly practiced in varying degrees, at different times, in different places. Perhaps the easiest and most common way to complicate geographic data production is to turn off the location services/GPS feature of one's smartphone. Location can still be tracked, of course, but typically by indirect, less accurate methods. For example, applications and the phone operating system itself will attempt to estimate location from local cell phone towers and Wi-Fi networks.

While many do engage in active resistance at times, it remains less common than simple acceptance for several reasons. First, resistance entails a much higher degree of motivation and engagement. Many motivations may spur an act of resistance, from disgust with the latest data scandal to political or professional necessity for privacy, as between protest organizers or journalists and their sources. But in each case, active resistance requires action: we must choose to purposefully depart the path of least resistance in order to (attempt to) reassert some control over our data. Furthermore, active resistance tactics are necessarily specific to technologies and their existing flaws and glitches. Just as terms of service and data generation processes are constantly shifting in response to these flaws, so must efforts to resist or subvert them. Today, bandanas or surgical masks are a decreasingly effective means of evading facial recognition algorithms. A decade ago, identifying someone in surveillance footage required a human observer. A decade before that, it was possible to avoid CCTV through careful travel in many cities.[5] As new technologies suffuse our daily practices, the cracks in which we may leverage our resistance appear and recede from view.

Regardless of the specific technology, forms of active resistance tend to involve complicating the production of data or obfuscating those data to make them less precise or meaningful, or some combination of the two. These acts can take many different forms, from the simple to the highly technical. At one end, supplying fake, fabricated information, like providing a false postal code when registering for a grocery store loyalty program or giving the email address of a politician you don't support when asked for one to receive a slight discount, are common practices that don't require additional technical familiarity expertise. Similarly, in many places people can wear full or partial facial coverings when attending a protest or otherwise engaged in activities that fall under increased surveillance. More technically complex measures include the use of end-to-end cryptography for email communications, virtual private networks for web browsing, data poisoning by providing a deluge of meaningless data through an extension like AdNauseam, and spoofing GPS readings on a smartphone to produce false location information.

Privacy advocates continue to release and support applications which engage in a variety of these practices. At time of writing, ProtonMail, run by Proton Technologies AG and with servers located in Switzerland, outside both EU and US jurisdiction, offers free encrypted email. The Tor network, run by the Tor Project, Inc. out of Massachusetts, uses a variety of techniques to attempt to preserve anonymity online.

Despite these best efforts, there are two flaws in these tactics of active resistance. First, they continue to rely upon the acts of individuals to step outside norms of use. Tinder requires Facebook to function, Facebook requires a verified email account, and so on. Individuals can step outside these systems to *escape*, or they can tinker at the edges, perhaps supplying a fake birth date, not using an email address tied to their name, or cleverly obfuscating facial images of themselves.

Second, individual acts of resistance, even when aggregated, run into an ongoing arms race between those who would extract data from users and those who seek to resist it. The existence of and resistance to timing attacks upon the Tor network demonstrates the limits of this approach at present. Without diving into the technical details, Tor can best be understood as working as an intermediary between a user and a website. Patrick O'Neill, writing for *Daily Dot* explains it succinctly:

A user fires up the client and connects to the network through what's called an entry node. To reach a website anonymously, the user's Internet traffic is then passed encrypted through a so-called middle relay and then an exit relay (and back again). That user-relay connection is called a circuit. The website on the receiving end doesn't know who is visiting, only that a faceless Tor user has connected. An eavesdropper shouldn't be able to know who the Tor user is either, thanks to the encrypted traffic being routed through 6,000 nodes in the network.

(O'Neill, 2020)

But, he goes on to illustrate a well-known flaw in the system. If an attacker has a system large enough to control *both* the entry and exit relays, then no matter how many nodes exist within the circuit, anonymity is compromised. This requires immense surveillance and computational power and is likely possible only for large-scale security agencies, like the US National Security Agency or China's Third Department of the People's Liberation Army's General Staff Department (3PLA). Such agencies exist and are, in many cases, exactly the types of surveillance one might wish to resist.

On a smaller scale and less technical in nature, many web browsers now offer features to block ads and the cookies used for tracking web use across sites. For several years, such efforts were broadly effective means of avoiding the annoyance of ads and making it more difficult for trackers to identify a single user across the web. More recently, many sites now detect the presence of an ad-blocker and ask or require users to disable it. Of course, ad-blocking tools are already beginning to be redesigned to be undetectable by such sites ... and so on. This is yet another example of how this kind of resistance runs athwart ongoing, asymmetric relations between technology users and creators. The individual subject, particularly the individual user, is not well positioned to resist the systemic efforts of the technological apparatus in which they live, one which mediates norms, both social and professional.

MAKING PRESENT

In contrast to tactics of active resistance which attempt to forestall the production of data, prevent their extraction, and/or limit their utility, *making present* refers to tactics which attempt to make the mechanisms of data production, extraction, and analysis legible and understandable. These methods counteract technology companies' efforts to elide such systems from consideration. Most often employed by artists, journalists, activists, and scholars, tactics of making present render unseen data, analysis, and even infrastructures visible to wide audiences, usually with the intent of opening such systems up to a wider public debate of their costs and consequences. Due to the intentionally communicative nature of this tactic, it has the most plentiful published examples to point to. Furthermore, the range of practitioners and their motivations engage a wealth of diverse audiences from highly theoretically oriented academic research to a casually engaged general public.

As we've already seen, perhaps the most common kind of making present is the work of journalists who raise questions and concerns around data hacks, data collection, and/or legal proceedings involving companies' and law enforcement's use of geographic data. While these exposés too often take the form of breathless and ahistorical recountings or reduce responsibility to the individual consumer, popular media remain a powerful way to impact public discourse, capable of reaching millions of people in a way that few scholarly articles do.

One powerful way to make things present is to call attention to the physical infrastructure of data in everyday life. Perhaps the best example of this is the work of artist-activist Ingrid Burrington on identifying, mapping, and visualizing data centers and major data cables. Doing so renders visible and understandable the material infrastructures on which the internet relies (Burrington 2016b). By travelling in person to data centers and creating guides to interpreting maintenance hole covers, Burrington's work calls these features out of banal invisibility and shows how close and embodied data infrastructure and associated state power are to each of us. In a similar move, the *Internet Atlas* attempts to present a "comprehensive repository of the physical internet" (http://internetatlas.org), but with a focus more towards academic research, rather than direct, embodied political engagement. More recently, and building directly on this and related works, both scholars (such as Lally et al. 2019; Levenda and Mahmoudi 2019; Hogan and Vonderau 2019; Nost 2020) and activist groups (such as Greenpeace's *Clicking Clean* report, Cook et al. 2016) have worked to make the intrinsic ties between our digital worlds and the material impacts they have upon our environment and climate more present.

In practice, data and infrastructure rely on one another, so other projects employ this connection to tie data imperatives to specific material geographies. The early work of Matt Zook, one-time state geographer of Kentucky, examined the geography of the internet as a geography of industry, wherein the relationality of space had material impacts (Zook 2005). One such example of the importance of "colocation" or "proximity hosting" in high-frequency trading (HFT) on stock exchanges. While HFT set-ups often involve significant sums spent on hardware and algorithms that shave fractions of a second off trade decision times, the physical, spatial relation between said hardware and the trading exchange is also of significant value. "Colocation," provided by many exchanges, situates that hardware within the same building and (ideally) on the same local network as the exchange, while "proximity hosting" refers to third party vendors that sell hosting valued in terms of its physical proximity to the exchange. In these cases, a few inches along a fiber optic cable may be worth hundreds of thousands of dollars. On a wider scale, *Regional Advantage: Culture and Competition in Silicon Valley and Route 128* by AnnaLee Saxenian explicates how the variegated geographies of peoples, places, environments, technologies, and *things* commingle to produce specific economic "hubs" of activity around certain industries, such as the San Francisco Bay Area and around Boston.

A number of artists use maps to better *make present* our current relations between data and selves. Some efforts have focused on the creation of maps of where surveillance cameras were located (see Figure 3.1), such as early efforts by the Institute for Applied Autonomy to generate travel paths that avoided as many CCTV cameras as possible in New York City in 2001. More recent collaborations, like The CCTV Map project, attempted to map surveillance cameras in and around London in 2012 (Zabou 2012). Other projects have attempted to map the paths taken by national security surveillance planes, such as Trevor Paglen's *Unmarked Planes and Hidden Geographies* project (2006) (see also the 2016 *Buzzfeed News* report by Peter Aldhous and Charles Seife, "Spies in the Skies," for a popular press account on similar phenomena). These efforts try to shift behaviors, or at least raise conscious thought on privacy and personal security, through the creation of visual images and multimedia installations that reveal what might otherwise remain invisible. How often does one really consider whether the plane overhead is conducting surveillance?

Although this is hardly a comprehensive survey of the work in this area, these examples suggest a focus on top-down surveillance, predominantly conducted by the state. These days, it's impossible to move across London or New York City without coming under the gaze of CCTV. It's also true that surveillance planes or drones are increasingly common both in the United States and elsewhere, as responses to Black Lives Matter protests made clear (Rose 2016; Aldhouse 2020). It is less obvious what happens when the tracking occurs "from the ground up" through the

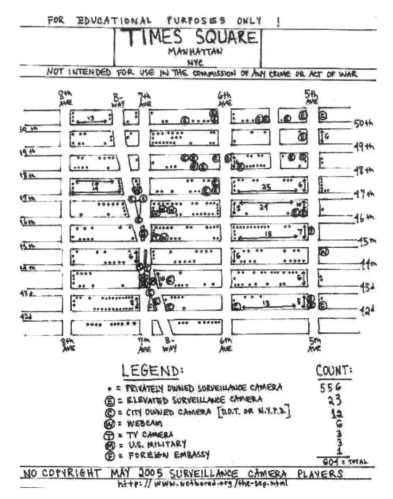

Figure 3.1 A map of surveillance cameras in Times Square, Manhattan, created by the Surveillance Camera Players in May 2005. This group drew direct inspiration from the Situationists as they contested the increasing surveillance of Manhattan after the attacks of September 11, 2001. We will return to these ideas in Chapter 4. www.notbored.org/times-square.html (last accessed July 2021). (No copyright)

willing acquiescence of a mobile device user. In the final entry in our typology, we explore the tactic (or impossibility) of *escape* before, in the next chapter, turning to potential strategies that may emerge through collective, rather than individual, actions with respect to spatial data.

ESCAPE

Throw out your phone. Use only cash. Wear masks everywhere …

While always wearing a mask may increasingly seem like a de facto norm for society in a post-COVID-19 world, the other ideas begin a list of possible steps to elude data creation and extraction. In many, even most, cases, efforts to escape are always partial and contingent, and many rest upon privilege and position within society. At times, escape is more performative than substantive. In this final section, we articulate some of the more common tactics individuals can use to escape while also noting the ultimate impossibility of total escape within modern global capitalism.

Jim uses a flip phone, and has for several years now. Apparently, so does Keanu Reeves (this is roughly where their similarities end), as do a number of other celebrities, including, according to CBC, Daniel Day-Lewis, Rihanna, and Kim Kardashian (Osler 2018). For many such people, these phones are simply supplements to other smartphones. It may not even be a conscious choice about data creation. For many celebrities, the "dumb" phone is an accessory that complements a "smart" one. The dumb phone acts as a more exclusive number only for voice calls or simply a marker of the ability to conspicuously consume by choice rather than necessity. For others, such as Jim and *allegedly* Keanu, the "dumb phone" is the only phone they carry. *Because they can.*

Neither has a job that forces them to be reachable by email at all times, much less requires them to carry a smartphone, as an Uber driver must. Neither operates in a situation where finding a map would be particularly difficult nor, as white men, is asking for directions terribly dangerous. Both have the capital necessary to survive if something goes awry, be it a broken-down car, missing a bus, or even simply deciding to walk into an Apple Store and purchase the latest model iPhone.[6] The ability to escape from the myriad of ways that smartphones create, mine, and extract data correlates directly with one's existing privilege within society.

For those who do only use a "dumb" phone, it's still not a full escape from location tracking. First, as long as the phone is connected to a network, location can still be tied to the closest cell phone tower (the one to which the phone is connected). Further, geolocation abilities in phones go back much further than many consumers realize. Starting in the mid-1990s, the US Government's Federal Communications Commission (FCC) initiated a phased implementation of Enhanced-911 that included precise geo-location of all mobile phones.[7] The policy required that by the end of 2005, 95 percent of all phones on a carrier's network must be able to be located within 125 meters at least 67 percent of the time. Carriers were allowed to determine how they would provide this location information, whether through the inclusion of GPS chips in phones, network-based triangulation (using cell towers), or some hybrid approach. Carriers that missed deadlines, including Sprint and US Cellular, were fined (Broache 2007). Likely due to the computational cost of network-based triangulation, the FCC estimated that more than 75 percent of all mobile devices had GPS chips by 2011 (FCC 2011). These mandates have continued to emerge and develop over time in both the US and other nations. China, for example, has required many commercial vehicles to use its navigational system, BeiDou, since 2013 (Wang 2013). In short, someone with a "dumb" phone, like Jim, is still be being tracked, but it reduces the accuracy, both in time and location, and it signals that they hold a status within society that does not require them to be always, instantly reachable.

Credit cards, with their time- and place-stamped records of spending, are another major source of spatial data tracking. Every purchase is logged and then (re)sold to vendors and advertisers to build a profile of predictable consumption habits, a smooth pattern of sales that can be targeted across platforms (email, mail, text, and so on). Of course, credit cards are also well established as to be nigh a necessity within systems of credit. How can one build the credit score necessary to secure a mortgage or even a lease without a credit card? These systems are intentional, and much has been written on the debt-based economy.

Again, the ability to escape exists at somewhat opposite ends of societal spectrums. On one side, if one possesses the capital and privilege to never need a loan, then a credit rating becomes far less important. On the other, where most of the world falls, to completely avoid systems of credit requires an abdication of many of the expectations and norms that allow one to function in society, such as leases for apartments or agricultural land, car loans, mortgages for houses, or loans for university study.

THE LIMITS OF THE INDIVIDUAL

Even before COVID-19, masks had become a flashpoint for resistance to surveillance in spaces like Hong Kong, where masks were banned in the fall of 2019 due to their use in pro-democracy protests (Elegant and McGregor 2019). Masks had served as a means of avoiding (or at least making more difficult) facial recognition surveillance by the state. A clothing firm, Adversarial Fashion, has developed an entire line of clothing meant to resist and escape efforts for surveillance; providing everything from hoodies meant to confuse license plate monitors to masks for facial recognition (http://adversarialfashion.com). In addition, many groups, like the Electronic Frontier Foundation, provide individuals with guides for how to safely secure their identity while protesting (or simply moving through daily life). And yet such active resistance and escape tactics rely upon individuals to engage in asymmetric acts against forces that sit very clearly on the other side of Andrejevic's big data divide (2014). Those with access to the data and means of analysis to draw conclusions (even spurious ones) from it are separated from those without. Barring a few exceptions, the majority of readers of this book, as well as its authors, do not have practical access to the mobile phone location data for time/location stamped photos of millions of individuals.

For those technologically empowered subjects sitting on their side of this divide, it doesn't matter if one individual, one time defeats a facial recognition algorithm. It matters that the algorithm is able to look through and process as many other facial images as it takes *until it is no longer defeated*. There is an invisible arms race between every individual engaged in resisting data extraction and the incredibly well vested firms and state organizations that seek such data and the analyses they facilitate.

So, then, for the rest of us sitting on the other side of the chasm, with limited access to the data and the systems necessary to analyze them, what is to be done? In the final two chapters, we argue for collectivized practices, strategies, that produce affinities between individuals. We suggest a move from the neoliberal subject to the collective group, a move done *through new systems of communication and data exchange*, and done in a way that supports a radical, liberatory politics, rather than data capitalism.

4
Contesting the Data Spectacle

No telephone. Write or turn up: 32 rue de la Montagne-Genevieve, Paris 5e.

(Internationale Situationniste no. 1, June 1958)

ASYMMETRIC POWER AND MOVING BEYOND THE INDIVIDUAL

At this point, it should be fairly clear that whether we like it or not, and even whether we're aware of it or not, the technologies of modern society are part of constant attempts to create, extract, and derive value from data produced through our daily actions. Many choose to accept the terms and conditions of use with little to no contestation, whether due to lack of consideration, calculated acceptance of costs and benefits, more immediate concerns, or because they have no choice. We aren't here to judge others on this: sometimes you just want Yu Xiang Qie Zi delivered and UberEats will give you 20 percent off if you tie your account to Facebook. But even those who choose to resist actively, passively, or attempt to escape data generating systems are ultimately caught in an asymmetric game where the vast majority of power is exercised by private, for-profit corporations for which users and their data are a means to an end. Our senses are flooded with data-driven spectacles, artificially shaping what can be known and what can be imagined in such contexts, and thereby foreclosing what might be done. Even when we look up from our devices and take out our earbuds, our lived environments are still suffused with the data collection and extraction processes that signal the core conditions of data capitalism. Once we are separated from our dataselves, they are fed back to us as affective sensations intended to produce specific actions, to sell specific commodities: Gregg's data spectacle (2015). But what happens to the data in the interim, how they are processed and classified, is not visible, much less available, to most of us. Technology companies separate individuals not only from their data, but also the tools with which to aggregate and analyze their data on their own terms (Andrejevic 2014; see also Wark 2004 and 2020). This separation of both data and the means to analyze them is not an accident, and it did not come about in a day. Such data dispossessions reflect the foundational social priorities that gave rise to these technology companies, and understanding that suggests some means by which these relationships may be challenged and changed.

In the last chapter, we focused on resistances practiced by individuals amidst their everyday lives. As actions of individuals, those tactics are inherently limited, with little prospect for broader systemic change (de Certeau 1984). One reason is that such actions are always-already incomplete, lacking the tools with which to confront and contest the structural power and associated strategies and technologies of large companies. Those tools, at work in Google search results, Facebook newsfeeds, and Apple's app store, allow those companies to operate at a different scale, combining and analyzing data from tens of millions or even billions of users. In most cases, these tools of data analysis, the very means by which data are rendered into actionable information, are closed to those outside the company. This is not just to retain the trade secrets of analysis, though that's part of the reason, nor is it simply a matter of expertise, though that's also part of the picture. Above all else, it is because those data processing tools are the means by which the company reaps the value of data, making the tools essential to the company's enterprise. Contesting these data relationships or refocusing them to more productive social ends therefore requires work not just on data themselves, but also how data are synthesized, analyzed, and classified on scales larger than what is possible for the self-valorizing individual consumer within the neoliberal order. This turf requires collective modes of resistance, a political praxis.

In this chapter, we'll first explore how the separation of ourselves from our data developed at Twitter. Twitter is an illustrative example of the separated, asymmetric nature of the relations between individual and platform-owning corporations. It also shows how the drive to colonize new moments of life as data in order to scale up profits developed and shaped what can be known across time and space. Second, building from the individual tactics described in the previous chapter and the theoretical works explored in Chapter 2, we propose and describe shifts towards collective, solidarity-building modes of resistance that could be employed to contest the data spectacle. These don't necessarily operate at the same scales as Twitter or other technology company platforms, but they nonetheless forge broader engagements for systemic change. We broadly classify these collective modes of resistance as *data regulation, data drifting, data détournement,* and *data strikes.*

But first, the curious case of Twitter's fire hose.

TWITTER: PUBLIC KNOWLEDGE VERSUS CAPITAL GROWTH

Twitter's revenues primarily come from two interrelated sources: advertising on Twitter's platform and selling users' data to better target advertising both on and beyond the platform. Direct advertising is the predominant source of income, making up approximately 84.5 percent of all revenue, or $682 million, in the first quarter of 2020. "Data licensing and other revenue," the only other category in Twitter's *Investor Fact Sheet,* accounted for an additional $125 million in revenue for that quarter (Twitter 2020b). While less than advertising sales, $125 million in three months is still quite a lot of money, especially for a revenue stream that emerged gradually over time. The case of how the sale of Twitter users' data by third parties became worth that much illustrates how data capitalism works to limit both what can be known and who can know it amidst an environment of ever-increasing profit seeking.

Less than three months after the first tweet in July 2006, Twitter released its first public API, which provided third parties access to Twitter content at a large scale in an effort to spur third-party app development to grow the platform and the company.[1] Early adoption of the API by third-party developers was strong, as a variety of applications began to make use of it to expand the functionalities of Twitter's services. This effectively allowed the company to externalize innovation and development of additional uses for Twitter, thereby attracting more users and leading existing users to spend more time with Twitter's services.

Researchers quickly began engaging with users' Tweets via the company's API as a new source of data. Efforts ranged from the Floating Sheep collective's playful Church vs. Beer map (Burn-Murdoch 2012) to more somber research to reveal hidden truths about the world, such as plaintive attempts to examine the impact of Twitter bots on the 2016 US presidential election. Among geographers, perhaps the best-known example of this was DOLLY (Digital OnLine Life and You), which built a massive repository of all geolocated tweets and provided initial indexing to make querying and analysis for other researchers easier. Under the technical direction of Ate Poorthuis, the project built a library of billions of tweets and made them available to researchers around the world.[2]

At first, Twitter's API provided free access to the entire continually renewed global stream of all user-generated Twitter data. This allowed third-party apps to copy and store all such data or subsets thereof, such as the attempt to copy and store every geolocated tweet using DOLLY. Then, after introducing a new API, Twitter began to segment access to its global stream of user-generated data. Over the subsequent years, the exact number of APIs and levels of access have changed repeatedly, but the company has always maintained a distinction between what's colloquially known as "sprinkler" (a.k.a. "garden hose") access, which provides a sample of tweets, and "fire hose" access, which provides full access to the entire stream. Initially, researchers could petition for access to higher "hoses" free of charge. In 2010, Twitter began making agreements with other companies to allow third-party reselling of Twitter data. Then in 2015, it acquired one of those companies, GNIP, and made GNIP the sole source for procuring large-scale Twitter data (Bryant 2015). While press release language focused on "inspir[ing] deeper analysis of tweets" (Van Grove 2010) and "promises to make data more accessible" (Bennett 2014), these actions by the company effectively locked guaranteed access above 1 percent of the tweet data stream behind paywalls.

At first glance, this is a niche, esoteric concern among socially privileged, technologically empowered stakeholders—"Oh no, researchers are having a harder time tracking down where people are talking about *Keeping Up with the Kardashians* more than *Terrace House!*"[3]—but the underlying impact on what can be known, by whom, and how they can know it is of much wider and substantial concern. This is textbook accumulation by dispossession, foundational to the processes of data capitalism. First, users generate the data through their everyday actions and Twitter leverages it to sell advertising.[4] Second, the company initially made the full extent of the data freely available to spur external innovation and grow the market. Third, when the external innovations were successful, it cut off the free data stream, enclosed the market, and added the revenue to its own profit margins. This *makes sense* for Twitter, as it is a publicly traded company required to file quarterly reports and show continued, sustained growth within a capitalist system.

And that's exactly the point. For researchers, what could be known through Twitter was initially structured by the technical characteristics of the company's APIs and subsequently by enclosure in pursuit of profit.[5] Today, Twitter's top-level API, "Enterprise" (or GNIP 2.0) is for "large-scale, well-resourced projects" (Twitter 2020a). Who gets to know what, when, and where is delimited by funding. Of course, within capitalism, this has been and likely will always be the case. Oxford and Harvard maintain libraries with more books and journals than either of our own home institutions.

DATA SPECTACLE AS A MECHANISM

Twitter's use of data to colonize and exploit everyday practices through analytics and opportunism is unsurprising amidst a capitalist system, but interpreting the company's actions is not enough. The point is to change these data-imbued relationships. In Chapter 2, we proposed the concept of the *spectacle* as one of the paths by which we might (re)assert our humanity within the technological relations of data capitalism. Here, we continue along that path, approaching data and associated analyses as a kind of spectacle, and therein plotting a course for collective resistance.

For Debord and the other Situationists, modern society was a series of spectacles, "a frozen moment of history in which it is impossible to experience real life or actively participate in the construction of the lived world" (Plant 1992, 1)—human experiences separated from each of us and fed back to us as commodified *images*. As stand-ins for actual human experiences, these images shape our understandings of the world and each other, both defining what is, as well as delimiting what we can imagine could or might be. The modern spectacle represented "the autocratic reign of the market economy which had acceded to an irresponsible sovereignty, and the totality of new techniques of government which accompanied this reign," a world controlled by images in pursuit of profit, defining and delimiting what is and what might be (Debord 1998, 2).

This approach draws from Walter Benjamin's observation that the alienation involved in the material production of society, its base, was seeping into its superstructure as capital found ways to exploit culture:[6] "[T]he dialectic of these conditions of production is evident in the superstructure, no less than in the economy" (Benjamin 2008[1936], 19). Debord develops this further: "The spectacle is not a collection of images; it is a social relation between people that is mediated by images" (Debord 1967, Thesis 4). Within spectacular society, time, culture, and social relationships are stripped from actual lives and cast as a disjointed, perpetual present of consumptive, commodified images, such that individuals are alienated "from their own experiences, emotions, creativity, and desire" (Plant 1992, 1). The spectacle is not just what is desired, but all that can be desired: "The spectacle turns the goods into The Good" (Wark 2013, 5).

The idea of the spectacle, particularly as a totalizing system in which all life occurs, is not without its critics. We've noted elsewhere that:

> the totality of the spectacle overstates nuances of lived experience and therefore weakens its conceptual utility, reducing it to an obvious intellectual fetish by which critical theorists may toss water balloons at the armoured tanks of capitalist modernity.
>
> (Thatcher and Dalton 2017, 137)

We argue that these views mistake "a totalizing tendency for a static totality" (Thatcher and Dalton 2017, 137). The spectacle seeks to colonize all aspects of life in all places and at all times by separating the *possible* from the *permitted* (Debord 1967, Thesis 25), but that doesn't mean it succeeds. As spectacle colonizes everyday lives, slipping into daily routines such as commuting or picking up the kids from daycare, it appears as targeted genre audio entertainment while you drive or a calendar that accounts for traffic delays. While siphoning off our data and feeding parts of them back to us, it shapes and optimizes the status quo, a more pleasurable commute or better optimized travel for daycare pickup. Spectacle will not facilitate more fundamental change, even if it is otherwise possible, such as different work patterns or paid family leave, unless it can find some way to capitalize on them as well. Alternatives must leverage that divide between what is actually possible versus what actions, beliefs, and imaginaries the spectacle attempts to limit us to.

SPECTACLE AS MORE THAN INDIVIDUAL

Such a conception of data, daily life, and what can be imagined runs the risk of being too centered on a single concept. If we're trying to resist the relations of data technologies today, why not center upon the work of Marx? Or Weber? Or Gramsci? Or why not abandon the dead white men and structure resistance around the ideas of Audre Lorde, bell hooks, Donna Haraway, Ruha Benjamin, Catherine D'Ignazio, and others? Without in any way overwriting such work and while encouraging our readers to seek out and follow the scholars and practitioners they find most useful, we offer one set of potential strategies for resistance focused on building solidarities in the face of asymmetric power relations. We frame the problem in terms of the spectacle for two intertwined reasons, while drawing inspiration, particularly in specific modes of resistance, from all of the above voices.

First, the idea of the spectacle is predicated upon dynamic communication and industrial technologies, but it also recognizes that the most effective ways to resist go through those very same technologies (Debord 1967, Thesis 24). For the Situationist International, the milieu of relations between the human and the technical was the site for resistance, for thinking and acting through alternatives to the given world (Wark 2011). Finding instances of that gap between the possible and the permitted and then leveraging them will inherently take place in the contemporary technological context and utilize technical means. Two examples include their experiments with détournement and the dérive, both of which we'll engage with in detail later in this chapter.

Second, and closely related, media theorist and scholar of the Situationist International McKenzie Wark suggests that the Situationists mark:

> the last of the historic avant-gardes. As such, they are something of a heretical formation within modernist culture, cross-pollinated with Marxism, and who proposed innovations not only in critical theory *but in organization, everyday life, and communication as well.*
>
> (Galloway et al. 2014, 158; emphasis ours)

It is at this confluence where new spatial data technologies work. This is where those technologies structure social organization, everyday experiences, and how we communicate with each other and ourselves. As such, the Situationists offer a unique opportunity for response, an untaken off-ramp from the expressway of the modern capitalist world (Wark 2011).

The question of why the Situationists keep returning is hardly a new one. Wark notes that "They stand in for all that up-to-date intellectual types think they have outgrown, and yet somehow the Situationists refuse to be left behind" (Wark 2013, 13). Derided even during their existence, there is a power behind their goal to "be at war with the whole world lightheartedly" (Debord, in Wark 2013, 15). Their refusal to live in commodified "dead" time and injunction to constantly seek pleasure in daily life are powerful ones that have echoed through the decades, even if their antics and infightings have made them a ripe target for mockery.

CONTESTATIONS

How can existing relations of data be resisted and changed to forge more equitable societies and empowering uses of geographic technologies? The measures of active resistance, making present, and escape we've covered thus far are small in scale. They make a difference on that scale, but are limited in their ability to provoke systemic change. Confronting and altering the data spectacle we face today requires larger actions, collective modes of resistance.

We outline four types of collective actions which hold promise for realizing systemic change, whether singly or in combination: data regulation, data dérive, data détournement, and data strikes. As with the typology of individualized tactics, our scope here is not to comprehensively engage every existing instance or case. Rather, our intent is to survey a range of promising possibilities and grounds for potential solidarities and empowerment through, rather than somehow against, new mobile and spatial technologies. With each mode of resistance, we seek to find ways by which these technologies can be made to *speak for* people in ways of their own choosing.

Data Regulation

Regulatory measures meant to protect personal data, including geographic data, have proliferated across multiple continents in recent years. The most prominent example is the European Union's General Data Protection Regulation (GDPR), which places legally enforceable limits on the collection, storage, processing, and transfer of individualized data. It also specifies non-anonymous or individually focused location information as a form of protected personal data within the regulatory terms. Thus, location data processing and analysis also fall under GDPR rules (Intersoft Consulting 2016). In practice, the GDPR requires active, informed consent to collect location data and combine them with other streams for targeted, non-anonymous purposes. This is very promising, and when fully enforced, changes the landscape of location data collection and how such data are used. Such regulation can limit the range of potential actions for major data companies, shaping the very design and coding of data-driven services to be more limited and responsible with users' data. The GDPR also shapes and delimits modes of data collection and handling in practice, normalizing such actions for whole populations, not just those invested in this topic. GDPR-compliant practices have become regular, everyday, even expected by data professionals and users alike across the European Union. Beyond the EU's borders, the GDPR has spillover effects as users in other jurisdictions utilize services hosted in the EU or built to GDPR specifications. As residents of the United States, Jim and Craig regularly encounter GDPR-compliant terms in the services we use. Finally, regulation such as the GDPR has a role to play in setting conditions favorable for other collective resistances.

For all these benefits, the GDPR's full scope, and thus its effects, are still under debate. As of this writing, the actual degree of enforcement is still being ironed out in court, both by technology companies and through class-action lawsuits from users. Even as these debates play out, the GDPR acts as an inspiration for similar regulations elsewhere. The new California Consumer Privacy Act (CCPA) imposes comparable restrictions, as do regulations in countries like Brazil, South Korea, and Japan. There is also discussion of similar legislation in the US federal government and some other countries.

Beyond the specific regulation of users' data, there are also increasing signs in multiple countries of antitrust regulatory actions against major technology companies, including Google and Facebook. The possible results of these actions are not just financial; an antitrust ruling could force a major technology firm to break up, and the subsequent implications for users' control of their data would be complex. Less centralization at a single company would make it harder to combine multiple data streams, which could make data somewhat less valuable, and thereby de-incentivize data collection to some extent, particularly in a GDPR-type environment that restricts data transfers between firms. Moreover, multiple smaller companies would likely be easier to regulate under a GDPR-style regime as they would have fewer resources to fight or find ways to circumvent the rules. At the same time, the multiple companies coming out of a breakup could proliferate the number of companies getting into the location data market or establishing their own data streams, complicating the experience of those attempting to better control their data. Furthermore, antitrust actions, even resulting in the breakup of major companies, do nothing to provide users tools to better analyze their own data for themselves, much less build solidarity between them.

Even with the promise of the GDPR and similar laws, *regulation alone will not save us*. It's not that regulation never works (it clearly can), but that regulation on its own is insufficient, even as it is the *only* systemic or collective form of resistance to current data regimes widely discussed in the popular press. While regulation has an increasingly promising track record in providing data subjects better control over their data, we need to keep in mind the limitations of what it can do.

Regulation of current data capitalism faces two broad limitations, one political, one cultural: First, regulations tend towards maintenance of the status quo of the spectacle. Structural change generally is not the intention so much as maintaining existing relations in line with contextual social standards and the demands of weighty political and financial stakeholders. Because formal regulatory policy inherently operates through governmental means, it is thus subject to disproportionately powerful corporate stakeholders. Technology companies employ armies of lobbyists and truckloads of campaign donations. In the United States, Google, Amazon, Microsoft, Facebook, and even TikTok all had lobbyists making *The Hill*'s 2020 "Top Lobbyists" list (*Hill* Staff 2020). As a result, getting substantial regulations passed into law is difficult. Moreover, once law, there is always the risk of regulatory capture or evasion. Such corporate strategies can hollow out a law, rendering it meaningless, such as through revised terms of service or arguing that a law simply does not apply to them, as Uber is notorious for doing. In extreme cases, powerful stakeholders may even reverse the law, say if a different political party takes power or through massive spending on a voter proposition. For example, in 2020, California Assembly Bill 5 took effect, ensuring basic labor protections such as minimum wage and sick leave for gig-economy workers. Later that same year, firms like Uber, Lyft, and DoorDash successfully collectively spent over $185 million on California Ballot Proposition 22,[7] which removed those protections, reclassifying gig workers as "independent contractors" (O'Brien 2020).

Second, regulations are governmental, which is good for crafting policy, but less effective in shaping culture and associated social change. The data spectacle works not only through economic and legal means, but also how data subjects see and understand the world and how they interact with one another. This cultural realm is where the limits of imagination and social acceptability are set, defining not only what is reasonable common sense, but also identifying problems that need to be dealt with. Cultural ideas of what conditions people should or will accept and endure don't tend to come from policy or regulations, nor do organized political movements that initiate structural change. As social movements have shown throughout history, these sorts of changes do not simply happen or begin through passage of a law. Effective regulations can shape everyday practice, as with actions that conform to GDPR rules, but there must be some social incentive to create and maintain a law like the GDPR in the first place.

Taking control over our data is far too important to be left exclusively to policymakers. More equitable data relations require cultural work that also falls outside the realm of governance and regulation. For geographic data, this means turning to additional collective modes of resistance that, while weird, are more creative and exploratory, working through cultural and grassroots means to impact how data subjects see, understand, and engage the world.

Data Dérive (Drifting)

Moving beyond regulation, we first turn to drifting, an exploratory practice to better understand individual and collective data regimes amidst the cultural geographies of daily life. Dérive (in English, "drifting") is a way of being aware of and learning about the structure and possibilities of a space while moving through it in a semi-intentional, exploratory manner, even if it is already familiar. This psychogeographic method provides reflection on the spaces, places, and practices of everyday life with the intention of identifying promising sites and modes for change, both social and material. In that way, drifting within the context of geographic data provides a promising way to understand and develop sites of collective resistance to current data regimes.

The concept of a dérive was first developed by the Letterist International (soon to be re-founded as the Situationist International) in response to the plans to redevelop Paris in the 1950s to include high-speed expressways. Discontent with this trajectory led a small group of leftist artists and activists to investigate and promote the pedestrian experience of the city's geographic structure:

The world we live in, and beginning with its material décor, is discovered to be narrower by the day ... this world governs our way of being, and it grinds us down. It is only from its rearrangement, or more precisely its sundering, that any possibility of organizing a superior way of life will emerge.

(Khatib 1958)

Following in the literal footsteps of Walter Benjamin's flâneur, they had the then revolutionary—today, utterly conventional—vision of an urban form that centered on the experience of pedestrians. As they walked, that experience could render the entire city as a single work of art (Chtcheglov 1953). Building from concepts found in Benjamin's *The Work of Art in the Age of Its Technological Reproducibility* (2008[1936]), the social significance of such art was not simply aesthetic beauty, it was political, and in the Situationists' case, revolutionary (Debord 1955 and 1956). While they were uniformly ignored by the authorities in the 1950s redevelopment of Paris, this idea was not as far-fetched as it may seem to modern eyes given the amount of revolutionary activity on Paris' streets for the previous 200 years. In fact, parts of it were borne out in the later 1968 rebellion, in which the Situationists played a starring role.

While urban form from the perspective of a pedestrian may seem run-of-the-mill in an era that valorizes "20-minute neighborhoods," the Situationists were responding to dominant, modernized views of the ideal city in that period.[8] Perhaps the paradigmatic example of such visions would be Le Corbusier's vision of whole planned, uniform, sanitized cities viewed from a distance (Figure 4.1). In contrast, the Situationist psychogeographic perspective was that of someone actually living in the city as it already existed. Specifically, they conceptualized the material city itself as a series of interconnected places they called "unities of ambiance," both planned and unintentional, that a pedestrian would experience emotionally and would be apparent in their behavior. These unities of ambiance were connected to one another by walkable streets, and that network structure literally encouraged the movement of people within and through them to particular streets and other unities, where they would feel and act, for good or ill, in ways befitting that place.

The dérive was a way of identifying unities of ambiance and the geographic network of connections and flows between them. It helped serve as the on-the-ground data collection behind maps such as *The Naked City*, *The Psychogeographic Guide of Paris*, and the related collage *Life Continues to Be Free and Easy* (Figure 4.2). In their thinking, by better understanding how the structure of a city worked, both as a material

Figure 4.1   Le Corbusier's Plan Voisin model for the redevelopment of Paris, displayed at the Nouveau Esprit Pavilion in 1925, is the sort of vision the Situationists we working against. (Wikimedia commons/SiefkinDR, licensed under Creative Commons Attribution-Share Alike 4.0 International, https://creativecommons.org/licenses/by-sa/4.0/, last accessed July 2021)

environment and as human experiences and behaviors in it, the urban geography could be better leveraged to realize revolutionary outcomes.

In practice, a Situationist dérive had no specific destination or planned route. Rather, it involved allowing the city structure and architecture to serve as a guide. It was similar to the urban walks of a flâneur, but the intention was revolutionary potential, not leisurely, refined consumer capitalism. Furthermore, the Situationist dérive was not random wandering. Rather, there was a specific method, a brutal commitment to a praxis of attempting to uncover the otherwise hidden structure and movements of the city directly. A drift could last a few hours or several days, and those performing it would record their experiences (Debord 1956). Abdelhafid Khatib provided the clearest account in his description of a dérive of the Les Halles district:

Considered from the viewpoint of the unity of ambiance, [Les Halles] differs only slightly from its official limits, and principally from an extremely large encroachment on the second arrondissement to the

Figure 4.2   Guy Debord's *Life Continues to Be Free and Easy* (1959).
This piece, sent as gift (potlatch) to Constant Nieuwenhuys, features
a collage of images placed upon a portion of the iconic The Naked
City map. Simon Sadler uses it for the frontispiece of his book *The
Situationist City* (2010), noting that "[i]ts layering of allusions—to
colonialism, war, urbanism, situationist 'psychogeography' and
playfulness—was dizzying." The original is preserved at the Rijksbureau
voor Kunsthistorische Documentatie, The Hague. (No copyright)

north. We observe the following boundaries: the Rue Saint-Denis to
the east [and so on] ....

   ... The architecture of the streets, and the changing décor which
enriches them every night, can give the impression that Les Halles is
a quarter that is difficult to penetrate. It is true that during the period
of nocturnal activity the logjam of lorries, the barricades of panniers,
the movement of workers with their mechanical or hand barrows,
prevents access to cars and almost constantly obliges the pedestrian to
alter his route (thus enormously favoring the circular anti-dérive) ....

   ... The essential feature of the urbanism of Les Halles is the mobile
aspect of pattern of lines of communication, having to do with the

different barriers and the temporary constructions which intervene by the hour on the public thoroughfare. The separated zones of ambiances, which remain strongly connected, converge in the one place: the Place des Deux-Ecus and the Bourse du Commerce (Rue de Viarme) complex.

(Khatib 1958)

For all their best attempts, this original Situationist form of drifting suffered from a serious conceptual failing. In their formulation, unities of ambiance were real, material things experienced in the same way by everyone, but as anyone who has ever walked through a city might realize, such drifting was highly subjective. They went out looking for unities of ambiance, but what they found as much reflected their own gendered privileges, racialized standpoints, and critical theory-informed perspectives as the material form of the city itself. In fact, an editorial note states that Abdelhafid Khatib was repeatedly harassed by police and arrested while drifting in Les Halles because at the time, North African men were forbidden from the streets at night (Khatib 1958). Despite this conceptual failing, parts of the Situationists' psychogeographic work now appear prophetic. Today, much urban design focuses on the experiences of pedestrians, and in 2012, Paris permanently closed the expressway along the Seine, making that space walkable once again.

In a broader context, the dérive has served as a basis for many subsequent geographical engagements with urban life and continues to offer hope as a collective means to resist current social relations. To us, the most significant of these sprang from an unlikely context given the privileged perspectives of many of the original Situationists and the ancestral connection to the flâneur. Precarias a la Deriva ("Precarious Women Adrift") was a feminist activist collective founded in Madrid in 2002. Sparked by calls for a general strike that year, group members confronted a question: how could they as gig/temp workers, domestic caregivers, self-employed workers, and similar feminized, precarious laborers go on strike? Excluded from site-based, male-dominated formal unions, they had no access to support like strike pay or other structures of formalized solidarity. Thus, in their "First Stutterings," they asked themselves, "What is your strike?" (Precarias a la Deriva 2003b). *What would a strike for precarious, feminized (and often reproductive) labor involve, and what would winning look like?* Building on the Situationists' dérive, they developed their own form of drifting designed to facilitate self-care and mutual aid in the places of their everyday lives as a way to address the situated challenges of precarious, feminized work and the multiple places where it occurs:

> We opted for the method of the drift as a form of articulating this diffuse network of situations and experiences, producing a subjective cartography of the metropolis through our daily routes.
>
> (Precarias a la Deriva 2003b)

By focusing on precarious labor amidst everyday life, Precarias a la Deriva's drifts (see Figure 4.3) were more socially reflexive than those of the Situationist Internationale. Precarias refashioned the subjective nature of drifting into a strength: a way to better investigate, understand, and empathize with participants' situated subjectivities for the purpose of nurturing new, radicalized subjectivities in solidarity with one another:

> In our particular version, we opt to exchange the arbitrary wandering of the flaneur [sic], so particular to the bourgeois male subject with nothing pressing to do, for a situated drift which would move through the daily spaces of each one of us, while maintaining the tactic's multisensorial and open character. Thus the drift is converted into a moving interview, crossed through by the collective perception of the environment.
>
> (Precarias a la Deriva 2003b)

Each of Precarias' drifts tagged along with a precarious working woman, at times one of themselves, through the practices, especially multiple forms of work, in their daily lives, often across multiple locations and contexts because precarious labor frequently is not limited to a single workplace (Casas-Cortés 2014).

This practice allowed all the drifters, including the person at the center of that drift, to pause and reflect on the conditions of their (often hidden) labor, the spaces in which it occurred, and ways to cope, resist, and build support networks and solidarity with others in similar positions. As a geographical practice, drifting allowed participants to trace these movements in space and thereby identify intersections and convergences as sites for resistances and forms of striking, confronting the social and material structures that otherwise keep such women isolated, separated, and less able to resist. By walking with migrant workers and asking them

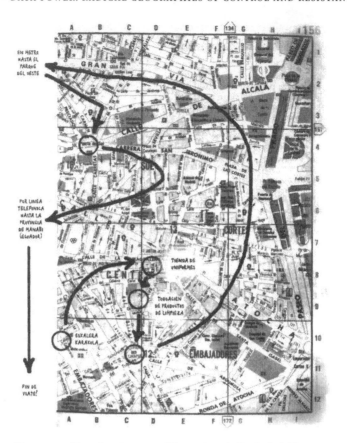

Figure 4.3 Map showing one of Precarias a la Deriva's drifts through the daily lives and practices of domestic workers in Madrid. (Used with creator's permission)

"What is your strike?", Precarias found those moments and places in daily life where said strikes could emerge, where solidarity could be built and leveraged in a situated manner (Precarias a la Deriva 2005).

Subsequently, other groups have adapted this Precarias-style drift in their own efforts. In North Carolina, members of the Counter-Cartographies Collective (3Cs) adapted it for the staff and graduate students of American universities and corporate research campuses. These drifts focused not only on the precarious labor conditions hidden by official university documents, but specifically emphasize how precarious labor plays an integral role in institutional knowledge production within higher education. What is and can be scientifically known and published, much less what knowledge is actionable for technical innovation, is frequently based on the precarious data-crunching labor of students, post-doctoral researchers, and adjunct faculty. Drifting provides a way to build solidarity with and among them in the spaces of their everyday lives (Casas-Cortés and Cobarrubias 2007).

Drifting, then, can play a profound role in building situated, placed solidarities and mutual understandings around places and data. It constitutes a collective mode of resistance to the data spectacle, and thereby data capitalism. Drifting confronts and re-frames the second part of Gregg's data spectacle, providing a way for users themselves to examine quantified representations of their dataselves and producing such geographic representations more on their own terms.

We propose engaging in data drifts, critically reflective examinations of the geographic data of everyday practices. A data drift repurposes data to better understand data geographies of daily life, their limits, and potential alternatives to the geographic visions of capitalized spectacle. Building on the original definition of dérive provides a place to start.

A data drift is a method of exploring the structures and possibilities of geographic data by moving through them, both physically and digitally. Doing so involves a mindset open to learning about the structures of such data, their connections with the social and material world, and the inherent limitations to such arrangements. This may involve spaces and data that are already familiar, so it requires careful attention to things and actions that are normalized or hiding in plain sight. Where the quantified self movement encourages self-tracking for individualized improvement, a data drift necessarily involves critical reflection and evaluation. It involves finding moments of solidarity with others, not a faster 10 kilometer running time. Data drifters can go where the data takes them through everyday actions, and thereby better understand themselves and those shaped by similar data processes.

Data drifts are directly inspired by the Situationists' dérive by focusing on the geographies of everyday practices. These are spaces and places from the perspectives and actions of those living in and moving through them, rather than an externally empowered view through digital data with a supposed global scope. It embraces the situated, reflective nature of Precarias a la Deriva's drifts, embracing the situated considerations of economic relations, cultural context, and particularly the gendered subject positions of participants as a means to better understand and empathize, and thereby to build solidarity. Furthermore, Precarias' drifts emphasized the role of forms of labor that are overlooked because they have disappeared into plain view: reproductive labor, care work, and the frequently precarious relations of those involved. So 3Cs extended this to knowledge-producing labor, which while more valorized, is directly connected to the creation of data. Data production is labor, even if it is unpaid or paid only by providing a service, because data are valuable within data capitalism (Fuchs 2014). However, much like the labor Precarias focused upon, the labor of such data tends to be ignored or overlooked as it occurs in the banal moments of everyday life.

So how does one actually do a data drift, what does it mean in practice? At its heart, a data drift involves moving through everyday geographic data and/or space paying special, critical attention to how data are collected and structured, their limits, their effects, and above all sites or openings for different kinds of data practices and associated social relations of data. Like other forms of the dérive, it inherently depends on the situation, participant(s), and the type(s) of data involved. It may be literally stationary, as some of 3Cs drifts were, and entirely mediated through devices, employing maps and tables to explore the data in question. Alternatively, it may involve moving through the spaces in question with a device and drifting mindset, akin to the Situationists, or employing the drifting mindset while going through the geographic practices of daily life as Precarias a la Deriva did. Likewise, the number of people involved will depend on the situation and focus of the drift. As Debord observed and others have replicated, while it is possible to drift alone, groups of two to four people are often the most fruitful. This is not for greater objectivity or inter-rater reliability, but rather, as Precarias demonstrated, to best build interpersonal solidarity and trust. This is also a vital difference from regulation as a collective mode of resistance. Drifting does not attempt to work at the national or state/provincial scale. Its exploratory focus is everyday geographies and solidarities, though the geographic insights, openings, and promising sites for resistance may be further leveraged by other collective actions. A drift's duration may be a few hours, a day, or as the Situationists did, even a series of days. The exact time spent drifting is less important than reaching familiarity or saturation with the data geographies in question.

That does not mean there is no material praxis to follow. The two indispensable aspects of a data drift are first, data situated in a context through which to drift, and second, the critical, investigatory drifting approach. A data drift will likely involve multiple interconnected data sets, though a single sufficiently large data set may serve. Not all the data sets need be spatial, but we find it helps to have geographic data or some kind of location-based index to help keep the drift cohesive.[9]

What might be learned by revisiting or cross-visiting where our phones believe we have been? (See Figure 4.4.) The best type of data is focused on the tasks and actions of daily life. Examples could include accessing the location history of a phone or an account with a major data platform such as Google, Apple, or Facebook. Accessing these kinds of data can be assisted by data regulations that require companies to provide users with the data the company has on them. Data tangentially connected to locations may also be helpful, such as ratings/reviews of places or sites. Still other kinds of relevant data may be accessible through less consumer-oriented firms such as credit rating agencies, and government records of property, registrations, or legal proceedings. What can be learned by walking from the most to the least expensive house in a real estate database of one's home town? For both reflective and ethical reasons, much data ought to be the participant's own. Moreover, data drifting in a group

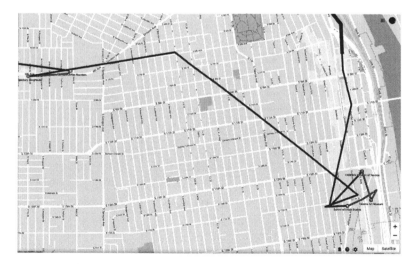

Figure 4.4 Walking route, Thursday, September 1, 2016. In late 2017, Jim recreated the path his phone had recorded for him that day, this time intentionally attempting to follow diagonal lines. This route brought him through new neighborhoods, past new businesses, and altered his understanding of the point-to-point relations between his work and his consumptive practices. (Created by the authors, data permissions in image)

with such intimate data ethically requires not only trust, but discussing and agreeing about expectations of confidentiality, ideally ahead of time.

Identifying the most useful or helpful data also depends on the second component: a critical approach to the situation. This involves both a reflexive mindset and purposeful actions, including how participants work with the data and what writing or making occurs as they think through the data. *Engage the data, play with them!* Move through the data physically and digitally, move through time with the data. Rank order items. Plot them on a map. Turn the map upside down. Run more advanced statistics if need be. *How frequently did drifters walk down a certain street, on what days?* Through that, pose and attempt to answer critical questions about what appears (*and what doesn't*), where those are (*and aren't*), and why? The data dérive is a means of opening for consideration the contours of the data spectacle, the very processes of data capitalism. Are there subjective unities of ambiance? Where are the contradictions and limitations? Where and what is a real material experience in light of and through the data? Where and what is spectacle? And above all, where are potential sites for solidarity, for disrupting the spectacular cycle of data creation, extraction, and analysis? Where are the moments, spaces, and times for resistance? In addressing these questions, we find that it helps to encode the experience of the drift to facilitate critical reflection. This may take the form of notes and subsequent written papers (as with the Situationists), making a map (as with the 3Cs), or even filming the process and creating a documentary, as Precarias a la Deriva did (Precarias a la Deriva 2003a).

For example, in one data drift in 2017, Jim examined the limitations and even patently absurd assumptions of his smart device's location history. It shows several locations he visited, including work and a coffee shop, but the vision of this spectacle also shows its farcical aspects. His apparent path through the day appears to have involved travelling only along Euclidean lines, diagonally across city blocks and straight through walls.

This sort of data drifting reveals not just the limitations of GPS waypoints, but also poses questions about what sites do not appear in this fun-house mirror reflection of a day's movements. Where is this system of data collection vulnerable? What sites and experiences are missing, not just because of technical glitches, but because they are not important to Google's data collection regime? Where is labor (of all kinds) happening? At what sites might different kinds of relationships be possible? It begins to ask: *Where is our strike?*

The drift is art, politics, and technology unified, the Situationist joyful struggle continued. Whether through personal reflection or solidarity, it holds value as a means to contest the data spectacle. That said, it is also inherently a small-scale, exploratory mode of research. What more can be accomplished at the sites identified in a drift? The Situationists and others provide more inspiration on this question, specifically with détournement and strikes.

Data Détournement

Détournement is a French term meaning rerouting, hijacking, and diversion, though not diversion as distraction in this context. The Situationists, particularly Asger Jorn, Jacqueline de Jong, René Viénet, Guy Debord, and Gil Wolman, conceptualized and practiced détournement as art, activism, or both simultaneously. According to the definitions published in the first issue of their bulletin, *Internationale Situationniste*, it is:

> Short for "détournement of preexisting aesthetic elements." The integration of present or past artistic productions into a superior construction of a milieu. In this sense there can be no situationist painting or music, but only a situationist use of those means. In a more elementary sense, détournement within the old cultural spheres is a method of propaganda, a method which reveals the wearing out and loss of importance of those spheres.
>
> (Situationist International 1958)

In practice, this involves modifying or disassembling/reassembling a piece of art or media for revolutionary purposes, typically in absurd or intentionally ironic ways that are more culturally savvy than traditional state propaganda or political campaign materials. In an age of mass media, much less today's easy media editing and internet distribution, this is a cheap way to make culturally salient political material, and thereby apply the capitalist spectacle against itself. Debord and Wolman cite Duchamp's mustache on the *Mona Lisa* and Berthold Brecht's cuts of Shakespeare's plays as precursors (Debord and Wolman 1956). A period Situationist example is *La dialectique peut-elle casser des briques?* ("Can

Figure 4.5 "Cat and Girl are Situationists," by Dorothy Gambrell. http://catandgirl.com/cat-and-girl-are-situationists/ (last accessed July 2021). (Shared with author's permission)

Dialectics Break Bricks?"), a dubbed version of the 1972 Hong Kong martial arts action film *Crush*, in which the audio has been replaced with a Marxist, Situationist narrative and references.[10] More recent examples include the many détourned works of Banksy, who integrates not just popular cultural iconography and political themes, but also location-based physical infrastructure, such as light poles, grates, and even the Israeli border wall into his street art.[11]

As opposed to just any piece of media modified for political purposes, Debord and Wolman outline two aspects of specifically Situationist détournement that are useful in forging modes of data resistance. In their *A User's Guide to Détournement*, they outline a double negation: first, calling out or removing the sense in which art serves as a commodity, then second, negating that negation, creating something that works towards positive political ends (Debord and Wolman 1956). This aligns with Walter Benjamin's earlier call for political purposes in art in his *The Work of Art in the Age of Its Technological Reproducibility* (2008), but the manner of Situationist détournement is more directly applicable to data. Détourning does not require producing wholly new things, but something more like a collage: adding elements or stitching different elements together so they mean something different, something more empowering.

Figure 4.6 A billboard in San Francisco détourned by the Billboard Liberation Front (BLF), a group devoted to "improving outdoor advertising since 1977." Its satirical press release, it reads: "'It's a win-win-win situation,' noted the BLF's DeCoverly. 'NSA gets the data it needs to keep America safe, telecom customers get free services, and AT&T makes a fortune. That kind of cooperation between the public and private sectors should serve as a model to all of us, and a harbinger of things to come.'" ("Billboard Liberation Front's take on AT&T Case" by dob, licensed under CC BY-SA 2.0, https://creativecommons.org/licenses/by/2.0/#, last accessed July 2021)

We propose a broadened and modified Situationist détournement inspired by these ideas, but that extends beyond art to data. Like art, data are powerful ingredients in forging new and different knowledges and understandings of the world. Moreover, it is harder to dismiss data on purely aesthetic grounds. While communicating data may involve aesthetic considerations, data can also facilitate material, evidence-based reasoning. This allows for more empowered subjects, provides grounds for building solidarity, and may even inform policy in wider contexts. To détourn data, we must first negate or at least set aside data's role as commodities. Then, those data may be repurposed or remixed, positively producing something new and different. Combining data in new ways can be both relatively easy and powerful—which, of course, is much of why it is so valuable as a commodity.

Data, especially geographic data, are useful outside commodity exchange. They have many applications for both broad strategic planning, such as environmental preservation and urban planning, as well as everyday actions, such as how to best travel to a friend's home. Furthermore, geographic data, whether a GPS coordinate or a mailing address, can serve as a shared index, a key with which to connect multiple different unrelated data sets in space, and often in time as well. For those attempting to use data to accumulate capital, this is how the individual that data capitalism can see is built: a credit card's billing address is tied to a cell phone number which in turn is tied to the subscription address for a loan servicer. A quantified profile emerges: this person buys speculative fiction audiobooks and owes student debt. As we'll see, data détournement can also leverage these characteristics through its double-inversion of relations.

In addition to focusing on data, the détournement we outline here differs from that of the Situationists in that we set aside their "laws on the use of détournement" (Debord and Wolman 1956). We argue that what is and isn't effective is contingent on cultural and social context. The right data for the topic, question, or purpose depends on contextual social processes and relations and their many manifestations such as language or the kinds of technical tools (or data!) available. Moreover, taking a page from Haraway (1988), it is vital for practitioners to proceed in a situated manner, aware not only of their own standing in relation to the data, but the people and living entities those data encode. Finally, the analyses and knowledge produced must be situated amidst these particular perspectives. For example, sousveillance data and even some appropriated surveillance data, such as police body cam footage, is ripe for data détournement.[12] Images, short videos, and memes from Black Lives Matter make clear the power of repurposed, remixed, data in building a popular movement.

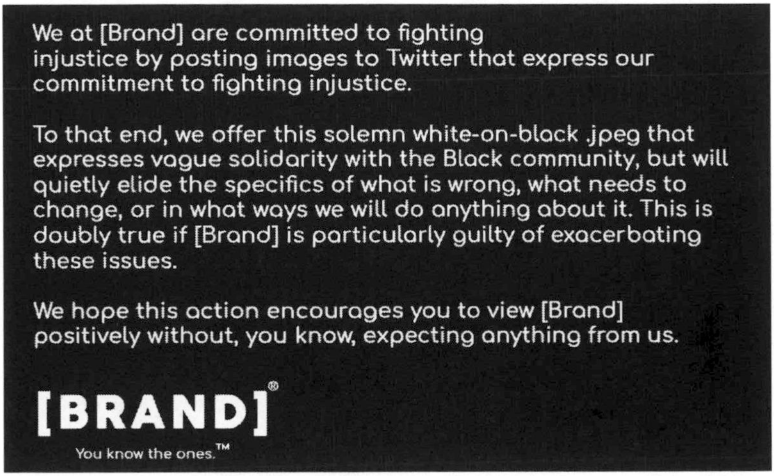

> We at [Brand] are committed to fighting injustice by posting images to Twitter that express our commitment to fighting injustice.
>
> To that end, we offer this solemn white-on-black .jpeg that expresses vague solidarity with the Black community, but will quietly elide the specifics of what is wrong, what needs to change, or in what ways we will do anything about it. This is doubly true if [Brand] is particularly guilty of exacerbating these issues.
>
> We hope this action encourages you to view [Brand] positively without, you know, expecting anything from us.
>
> **[BRAND]**®
> You know the ones.™

Figure 4.7 A détournement of corporate statements made in support of Black Lives Matter, created by Chris Franklin, (@Campster) and posted to Twitter on May 31, 2020. https://twitter.com/Campster/status/1267183124582215680/photo/1 (last accessed July 2021). (Shared with author's permission)

At a glance, data détournement resembles those creative hacking moments that produce innovative technical fixes or software coding practices that pull code from multiple sources such as GitHub repositories. Not all hacking and coding practices would qualify as data détournement, but some undoubtedly would. Hacking and coding projects that do not attempt to commodify the data involved and that instead creatively apply those data towards productive political or social outcomes can be understood as forms of data détournement. In addition to the powerful, political examples above, less serious, playful détournements of data also exist. For example, the Twitter account @Marxbot1 will respond to any tweets directed towards it with text generated by a Markov chain trained on Marx's works.[13] This is hardly a serious intervention, but a playful one that explores the entanglements of modern artificial intelligence and the works of Karl Marx.[14]

Among geographic data and practices, mapping provides fertile ground for data détournement. Professor John Pickles of the University of North Carolina at Chapel Hill argues that all mapmaking projects, at least as we understand maps today, are a sort of collage of data points and information, assembling multiple data sets and putting them together to serve a contextually defined intent or purpose (Pickles 2004). For example, even a relatively straightforward tourist guide map combines data of the location and characteristics of roads, political boundaries, topography, water bodies, protected lands, and many other themes. Even gas stations, McDonald's, and other corporate sponsors often find themselves in ostensibly state-created maps. Each of those kinds of data in a given area may come from a different source, such as government data sets or users' location histories. Only in the visual database of a map are all of these data sets assembled together in a manner useful for visual interpretation.

This is not to imply that all mapping is data détournement. For most of the last five centuries, mapmaking has tended to serve the purposes of large social institutions, such as governments and corporations, whose interests often do not align with people on the ground. Examples include colonial expansion, redlining to facilitate housing discrimination, and demographic profiling. But not all mapmaking facilitates such processes. Counter-mapping, for example, attempts and succeeds in flipping these relations: the people who have traditionally been on the receiving end take control of cartographic tools in order to make maps for and of themselves (Peluso 1995; Counter Cartographies Collective et al. 2012). As such, many counter-mapping initiatives can be described as data détournement. Its practitioners tend to work outside capitalist social relations, combining and remixing data to map in ways that explore alternative social or environmental possibilities and that can seek to make those possibilities reality (Dalton and Stallman 2018). In this way, counter-mapping projects can accomplish more than an exposé or the actions of a single person. They put data to work to positively initiate change rather than simply revealing their scandalous existence or trying to better control the data extracted from individuals.

Counter-mapping projects provide great examples of data détournement and what can be accomplished through it. Some kinds of counter-mapping attempt to educate readers, building interest, and possibly a movement, and therein influence cultural formations and policymaking. This is similar to the individual tactic of making present, but here operates collectively, both in the data/mapmaking practice and the broader context of a social movement. One powerful case of such counter-mapping as data détournement is Inside Airbnb. This small collective reveals the impact of Airbnb's gig-economy business model on housing prices as part of the push for more affordable housing in the United States and around the world. Specifically, the concern is that as more landlords list more houses and apartments on Airbnb's short-term rental listings, that housing is removed from the residential market, exacerbating housing shortages and driving up rents. Inside Airbnb confronts this by scraping all the listings from Airbnb's website in major world cities and mapping them. Then it goes on to estimate how often each listing is occupied, to indicate how much housing is removed from the residential market by neighborhood.

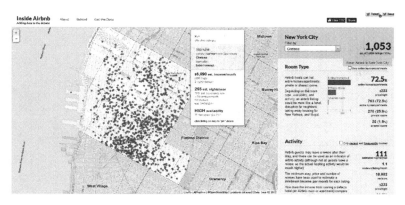

Figure 4.8 A screenshot of Inside Airbnb's web map, created by Murray Cox and Inside Airbnb. (Shared with author's permission)

These efforts have directly impacted policymaking aimed at better regulating the Airbnb company. In 2015, Airbnb released a data set of its New York City listings as part of a public relations campaign. By comparing that data to its own scraped copy, Inside Airbnb revealed that the company had removed over a thousand illegal or embarrassing listings before releasing the data set to the public. The company admitted to removing the listings, and the New York State legislature subsequently passed more restrictive regulations aimed at short-term rentals and Airbnb (Dalton 2020; New York State 2016). This episode is not just a clear case of détournement, it also demonstrates what data détournement can accomplish in the right situation. Inside Airbnb used the company's own data against it by utilizing the data not as a commodity, but as a political tool.

Other counter-mapping cases show how flexible data détournement can be in this sort of overall strategy. The data-driven organization Mapping Police Violence collects and maps every death at the hands of a law enforcement officer in the United States because there is no comprehensive governmental database of these deaths.[15] The Federal Bureau of Investigations keeps a list, but additions from other branches of government, including local police, are presently voluntary. By assembling all available information on deaths into a single national database, Mapping Police Violence and other allied organizations doing similar work show both the geographic distribution and racialized nature of killings by police. A similar worldwide case was Civic Media's 2015 map of tear gas use. Again, there is no official comprehensive data set, so Civic Media Hub assembled one to better illustrate the use of this "less than lethal" crowd control weapon. Perhaps the most prolific group employing this approach in the United States is the Anti-Eviction Mapping Project (AEMP), which was first launched to make the wave of evictions in San Francisco visible and legible. Over time, the collective has expanded to map a wide variety of topics and places in the struggle for affordable housing and communities facing gentrification (Graziani and Shi 2020). AEMP employs a huge variety of data sources, from the US Census and city records to court proceedings on evictions to stories told by people who were evicted, ultimately creating their own data through collaborations with allied organizations and communities.

Other counter-mapping groups that engage in data détournement focus on direct action. In many cases, these kinds of maps are protest tools, whether as a paper guide for protesting the Republican National

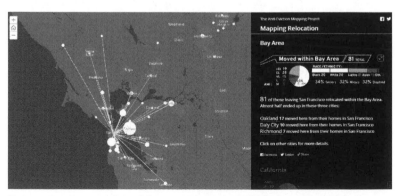

Figure 4.9   A screenshot of the Anti Eviction Mapping Project's Mapping Relocation map. (Shared with creators' permissions)

Convention in 2004 or later phone-based map applications. These kinds of actions can be very powerful on-site, but their impact tends to fade quickly in the continuing technological arms race of protest and policing. One influential early application was Sukey, which allowed student protestors in London to out-flank police efforts at kettling and to continue protesting based on geographic suggestions from app admins off-site.[16] More recently, pro-democracy protesters in Hong Kong used a crowdsourced, decentralized map application, HKmap.live, to better coordinate their responses to increasingly militarized police responses. The Chinese government deemed that map to be such a threat that they forced Apple to remove HKmap.live from the company's app store (He 2019), something the company would only do for noted white-supremacist app Parler after some of its users stormed the US Capitol building in an attempt to overturn a democratic election.[17]

Like drifting, data détournement is not suitable in all situations or all cases. In particular, those who use it need to be careful of privacy, given that such data aren't necessarily about themselves. Détourning some data about other people, even in groups, may not be ethically possible (Dalton and Stallman 2018). In contrast, the last data-derived collective mode of resistance we'll cover here starts with one's own, personal data.

Data Strikes

Strikes are perhaps the most conventional collective mode of resistance we address here, and yet the least developed in actual practices concerning data. By "strike," we mean an adapted version of a traditional workers' strike: denying labor, in this case data-producing practices, to extract concessions from management and ownership. In this case, the management and "ownership" are major data companies and those involved in capitalist pursuits through the acquisition, analysis, and trade of said data. A data strike involves coordinated withholding of data sufficient to endanger data companies' profits to force them to make changes in how they conduct their business and its impact on the lives of the people from whom they extract data. One user denying their labor is too minor for companies to notice, but a coordinated mass of users can spell serious trouble for such firms.

Admittedly, there is a gray area here between a data strike and a boycott. An actual data strike could borrow boycott-style elements, such as ways of organizing to define demands. Nevertheless, we use the word "strike" to emphasize the role of data-producing labor performed by so-called users and as a call-back to the efforts of Precarias a la Deriva and their inspiring question: "What is your strike?" Boycotts have a great history of success, from busing in Montgomery to ending South Africa's Apartheid to the labor behind Taco Bell's tomatoes, among others (Friedman 2017). Nevertheless, due to users' role in producing data value, they play a different role than the word "boycott" traditionally suggests. They create data, and the data are extracted from them, rather than simply acting as a consumer who pays money for goods or services such as riding a bus or purchasing a sweater. Professor of digital media and critic of capitalism Christian Fuchs describes in detail how everyday digital practices, in particular activity on social media, are forms of labor. Reading, much less posting on Facebook, searching for pizza on DoorDash, and using turn-by-turn Google Maps directions are all actions that directly or indirectly produce data. As we discussed earlier, data are valuable, which is why companies collect and use them for their own services and targeting of ads or sell them as a commodity to other companies. For our purposes here, the crucial point lies in the value produced through using those services, transforming that usage into a form of data-producing labor (Fuchs 2014). That, in turn, raises the possibility of a strike—that moment when enough people withhold their labor in concert with one another, thereby stopping or restricting the flow of valuable data enough to force a company to respond. Furthermore, casting users as data-producing workers provides a basis for building solidarity with other workers in different roles within the technology industries, such as Twitter interns, Uber drivers, Google coding contractors, Facebook content moderators, and even full-time Microsoft software developers.

Among users, withholding labor can take several forms. The most straightforward is users limiting or even stopping their usage of a data-driven service, thereby reducing not only that company's current business, but its potential growth through that user. Moreover, due to the socially networked nature of much digital data, users withholding labor have knock-on effects as it subsequently reduces the performance and thus the value of data of still-active users. Moreover, the very data services themselves can become a basis for building solidarity between strikers and maintaining the strike (Arrieta-Ibarra et al. 2018). Users can inflict still greater damage by not only withholding their current and future labor, but by deleting all of their past contributions as well, thereby leveraging the cumulative nature of the value of personal data. For example, Craig could not only stop using Facebook, he could delete his account and all associated data as well. This method comes with greater costs for both the company and the users, for it would make it harder to start using the service again afterward. Companies tend to make it difficult for a user to delete all of their data in a service, presumably to ease re-activation and because even the data of former users are still valuable. Thus, the potential of this mode of striking by deleting accounts depends in part on a regulatory environment that forces companies to actually delete a user's data when the user requests it. The EU's GDPR and California's CCPA include such requirements, indicating both the importance of such regulatory regimes and also their insufficiency without additional collective action. Finally, in certain cases, even just a well-coordinated threat of a data strike may prompt concessions from a company under the microscope of potential venture capital or Wall Street investors.

Even with these strengths, a data strike faces distinct challenges. First, withholding data from some companies is extremely difficult in today's society. Data brokers, credit rating companies, security contractors, analytics firms, and similar corporations have minimal "consumer"-facing services and tend to procure data from publicly available sources and other indirect, hard-to-avoid means. That's not to say a data strike against Acxiom, Equifax, or Palantir is impossible, but it will be more difficult than against Facebook or Twitter. Second, as with a traditional strike, there are costs for the strikers. A data strike means going without or finding alternatives to the digital tools and services of everyday life. That may be easy for Pokémon Go or even Facebook, but services like Gmail or online bill payment services would be rather more difficult to set aside for many. Some contractors and gig workers rely on cheap email, and their businesses are often closely connected to such accounts. In such cases, it may be necessary to minimize, rather than end, the use of a given company, for example to bring intentionality back into Uber Eats orders. Third, a strike requires coordination to get the word out, build solidarity, encourage adherence, and for negotiation. The everyday, contextual nature of data labor makes this difficult, but not impossible. Most scholarship on the potential for data strikes to date focuses on the idea of a union of users as the mechanism for organizing a strike and negotiating collectively (Arrieta-Ibarra et al. 2018; Posner and Weyl 2018; Vincent et al. 2019). Building such a union would undoubtedly be a challenge, though quite possibly a fruitful way to extract concessions from major technology firms. However, cases as diverse as the Situationists, the Student Nonviolent Coordinating Committee, Precarias a la Deriva, Occupy Wall Street, and Black Lives Matter indicate that a union is not necessary per se for a coordinated social movement to have significant effects. Moreover, strikers can employ other internet services to build solidarity during a strike as long as it isn't a strike against all data companies simultaneously. Combined, these factors mean that a data strike may be difficult, but no strike is easy.

What does a data strike look like? To date, data strike actions are rare, but both modeling and a few cases provide inspiration for what is possible. Research by Nicholas Vincent, Brent Hecht, and Shilad Sen attempts to model the impacts of a strike on algorithmic recommendation systems for movies, such as Netflix's suggested viewing. In their initial findings, a strike in which 30 percent of users deleted their accounts halved the performance of the recommendation system, and if 37.5 percent of users participated, the recommender algorithm's performance for non-striking users degraded the accuracy of such systems down to what they were in 1999. Moreover, due to the design of the recommendation algorithm, some demographics could exert an outsized impact in a strike. For example, fans of horror movies and users under the age of 18 could cause disproportionate damage to the system (Vincent et al. 2019).

In practice, data strikes and similar actions thus far are rare, but not unheard of. One case that hints at the possibilities of a data strike emerged in early 2021, when WhatsApp announced a new privacy policy allowing third-party businesses to store WhatsApp chat logs on the servers of WhatsApp's corporate owner, Facebook. Moreover, the update, appearing as a pop-up, was required in order to continue using the service. Many WhatsApp users reacted negatively, believing this to be a new violation by Facebook, which already had a shoddy reputation on privacy. In fact, WhatsApp had long shared some information with Facebook, such as user phone numbers, while it claimed to not share the contents of encrypted messages sent on the service. Nevertheless, there was a widespread public outcry, including celebrities tweeting to promote competing messaging services not owned by Facebook. Of those competitors, Signal became one of the most downloaded apps on the Android and iOS app stores overnight and its new-user verification system crashed under the load. Telegram, close behind, added 25 million new users in the first three days (Statt 2021). WhatsApp and

Facebook executives scrambled to contain the damage, postponing the new privacy policy for months to revamp the rollout, though it was ultimately still mandatory. While not a full data strike, this case provides some indication of just how fearful major technology companies are of losing their stream of user data.

#DeleteUber provides another promising case. It was a campaign, largely on social media, to stop using Uber services, primarily the rideshare/taxi service, because of objectionable actions by the company. The hashtag first appeared on January 28, 2017, when President Trump suddenly announced a new ban on travelers entering the United States from seven majority Muslim countries, sparking protests, most famously at New York's John F. Kennedy airport. Unionized taxi drivers in the city, in solidarity with the protests, stopped providing service to JFK, but Uber and Lyft continued to provide airport service and Uber removed surge pricing, making those rides cheaper. Many rideshare users, eventually numbering in the hundreds of thousands, reacted to the company's apparent acquiescence to the president's new policy by deactivating or uninstalling the Uber app from their phones and sharing that action on social media with the tag #DeleteUber. Although it was estimated that only 0.5 percent of Uber's active user base participated at the time, it grew into yet another negative narrative for the venture capital-backed company as it prepared for its initial public offering (Shen 2017). While it is impossible to isolate the full impact of the campaign from Uber's other scandals of the period, #DeleteUber was certainly a contributing factor to CEO Travis Kalanick's exit from the company in June of that year. Furthermore, during the campaign, Lyft's new app downloads outpaced Uber's for the first time, even as Lyft also continued to provide service to JFK and didn't remove surge pricing, meaning it likely profited more than Uber from airport service that night (Lee 2017). No strike is perfect.

A data strike is difficult because of the way data companies exploit not only formal labor, but also our reproductive labor practices, and even our play. They are part of our communications, our calendars, our navigation, and our shopping lists. Confronting that takes creativity. The flights from WhatsApp and #DeleteUber were spontaneous, decentralized, and short-lived, but still had some impact. Not all data strikes need follow this form. Some could incorporate tactics such as active resistance practices of blocking data collection and coordinate them in a more collective, concerted, campaign. What could be accomplished with more organization or clearer demands?

Precarias a la Deriva asked, "What is your strike?", and in so doing, powerfully demonstrated that the positive outcomes of a collective mode of resistance need not be a collective bargaining agreement or contract. Through their struggle, they built solidarity, mutual aid, and shared survival strategies, making their situation more livable *together*. Confronting data capitalism is much the same. Voluntary changes by major firms would be great, and actual, enforced regulations even better. But in practice, collective resistance to the relations of data will be realized in everyday life through methods such as drifting, détournement, strikes, and likely through other creative means impossible to imagine now. Initiating change is far too important to be left to policymakers and corporate boards. In Precarias a la Deriva's words:

> [T]he strike appears to us as an everyday and multiple practice: there will be those who propose transforming public space, converting spaces of consumption into places of encounter and play preparing a "reclaim the streets," those who suggest organizing a work stoppage in the hospital when the work conditions don't allow the nurses to take care of themselves as they deserve, those who decide to turn off their alarm clocks, call in sick and give herself a day off as a present, and those who prefer to join others in order to say "that's enough" to the clients that refuse to wear condoms ... there will be those who oppose the deportation of miners from the "refuge" centers where they work, those who dare—like the March 11th Victims' Association (la asociación de afectados 11M)—to bring care to political debate proposing measures and refusing utilizations of the situation by political parties, those who throw the apron out the window and ask why so much cleaning? And those who join forces in order to demand that they be cared for as quadriplegics and not as "poor things" to be pitied, as people without economic resources and not as stupid people, as immigrants without papers and not as potential delinquents, as autonomous persons and not as institutionalized dependents. There will be those who ...
>
> (Precarias a la Deriva 2005)

... put down their phones and share real, unmediated human contact. There will be those who together make their data their own. There will be those who ...

5
Our Data Are Us, So Make Them Ours

In their designs and assumptions, algorithms shape the world in which they're used. To decide whether to include or exclude a data input, or to weight one feature over another are not merely technical questions—they're also political propositions about what a society can and should be like.

(Amoore 2020b)

[Y]our scientists were so preoccupied with whether or not they could that they didn't stop to think if they should.

(Ian Malcolm, *Jurassic Park*, 1993)

In the midst of the COVID-19 pandemic, both technology boosters and many politicians turned towards spatial data as a potential savior. Given all that our phones can do, how could they not save us from this too? The promise of technology burned ever brighter as multiple governments rolled out applications that would use spatial data to track when, how, and if users came in contact with anyone with the virus. Of course, as is so often the case, these didn't work. Privacy concerns shunted some projects into the dustbin, lack of adoption sank others. Regardless of the reason, ultimately the applications *simply didn't work*. As we consider the pandemic, it's worth recognizing why these systems failed and will continue to do so into the future.

Yes, mobile phones track our daily movements at certain temporal and spatial scales, and yes, services on those phones facilitate shifts in how we move through and come to know the spaces and places of our lives. However, even as these data come to stand for us in a variety of algorithmic and data-based systems, the fundamental goal of these data are not to fully capture our experiences, but rather to render us calculable within systems of capitalist consumption. The temporal and spatial accuracy of these systems are designed to better target users for advertisements or predict major purchases, to decide when and if someone is ready for a mortgage (and at what rate), to evaluate potential job applicants … and none of these intrinsically capture the nature of intimate, physical contact by which a disease spreads.

Elsewhere, we have called this "the epistemological leap of big data": from a calculated representation of an individual meant to sell them socks to the full, lived experience of that individual's life (Dalton and Thatcher 2015). The individual our spatial data can see is the individual capital wants to see, an algorithmically sorted consumptive bracket able to be called forth in a predictable manner. This is not the full social life of humans, nor does it seek to be, and as such, attempting to force one to represent the other will inevitably fail.

COVID-19 gave the world an opportunity to confront this failure, to see the stark gulf that exists between the representation of an individual through their spatial data sorted through various algorithms, and the full serendipitous practices that make up daily life which constitute our selves. In this chapter, we tie together the threads woven throughout the book to present one means of seizing this opportunity, of making the data that are us into our own.

(NOT QUITE) AGAINST THE ALGORITHM

Unable to have students sit their A-level exams due to the pandemic,[1] the UK government turned to Ofqual, an exams regulator, to provide an algorithm to assign grades to pupils in England. Ofqual, in an unsurprising move, kept the majority of its algorithms' inner workings as proprietary secrets.

As anyone reading this far into the book can guess, a fiasco ensued. Without diving fully into the technical workings of Ofqual's algorithm, the crux of the issue revolved around how the algorithm used students' circumstances to automatically adjust grades. These circumstances included aspects outside the students' control, such as the achievement of students from their school in previous years. Ofqual's algorithm decided on the most likely highest and lowest grades for each class, then force-fitted students' scores across that range. No matter how hard a student had studied or how hard teachers had worked to improve a school, Ofqual's algorithm would not (and by design could not) take these factors into account. Predictably, students from disadvantaged areas were assigned lower scores. Ofqual also, naturally, removed students' ability to dispute grades.

At time of writing, the uproar continues to swell and the UK government has promised to reverse course and use an approach based upon individual teachers' estimations of likely scores for their pupils. Nevertheless, the damage is done. University acceptance letters have already been sent out. Oxford University, for example, has stated it is "unable to offer further places to state school applicants affected by the grading fiasco because of a cap on numbers imposed by the government" (Adams and Stewart 2020).

Regardless of how this issue is ultimately (not) resolved, Durham University Professor of Geography Louise Amoore writes that protests against it illustrate a shift in how society understands the danger of algorithms and data:

> Resistance to algorithms has often focused on issues such as data protection and privacy. The young people protesting against Ofqual's algorithm were challenging something different. They weren't focused on how their data might be used in the future, but how data had been actively used to change their futures. The potential pathways open to young people were reduced, limiting their life chances according to an oblique prediction.
>
> (Amoore 2020b)

This echoes the call found in Virginia Eubanks' *Automating Inequality* (2018) to heed how data and algorithms are used on the poor and non-white populations, not only for the sake of justice, but also because they already live in the future. How those without the means to resist are forced to accept data extraction and algorithmic governance is a stark template for how it will and already has crept into wider daily discourses, particularly in light of a pandemic that's best addressed through contact-tracing (tracking). The A-level fiasco is simply one more in a litany of examples of algorithms and the data through which they operate failing to recognize the particular in light of the general. As is so often the case, this burden fell disproportionately upon those who were already disadvantaged. If a student was from a poor school, they were likely to be graded downwards.

Yet we seek a different path than Amoore's claim that "'Ditch the algorithm' is the future of political protest." We instead call for a politics that resituates data and algorithms within the purview of liberatory experience and solidarity. Neither algorithms nor data are ever neutral, but nor do algorithms intrinsically "clos[e] off spaces for public challenges that are vital to democracy" (Amoore 2020b). Rather, we find that their current uses within discriminatory, capitalist, profit-seeking systems lead to that dark end. In these concluding moments, we summarize three key themes that begin to resituate spatial technologies within the radical praxis of the everyday.

INFORMED DAILY USE, A PRACTICE-BASED PERSPECTIVE

Ultimately, this book asks: what are the liberatory ideas and actions, the politics of emancipation, that can or might occur *with* new spatial technologies? In *An Essay on Liberation*, Marcuse (1969, 12) asks:

> Is it still necessary to repeat that science and technology are the great vehicles of liberation, and that it is only their use and restriction in the repressive society which makes them into vehicles of domination?"

First, it is necessary to acknowledge that the very premise of this question is not, nor has it ever really been, universally accepted. Whose science and whose technology matter not simply in the abstract, utopic sense of communal control, but also in the very ways that epistemology and ontology demarcate violence against alternative ways of knowing and being-in-the-world. There is much to be said about the (potential) incompatibility of western scientific rationality and other ways of knowing.[2] At the same time, such scientific rationality is in part built from other knowledges. As Professor of American Cultures Lisa Nakamura has eloquently argued, the very construction of these technologies has leveraged both the imagery and bodies of indigenous and other peoples (see, for example, Nakamura 2014).

We return to Marcuse's question with some sincerity, détourning it a bit, to ask how we might remake systems of technical exploitation into vehicles of liberation—not against the algorithm, but a politics *with* it. We opened this book by suggesting that we live within the ruins of a technocapitalist system that has denied our humanity and destroyed our environment, and throughout it we have demonstrated how that system continues to instantiate itself as perpetually new and forever unforeseen as a means of obscuring critical inquiry and avoiding meaningful regulation.

Where brilliant legal scholars, like Frank Pasquale (2015) and others, see a path forward through regulation and legislation, we see that as another utopic approach to a declaration of "science and technology" as inherently liberatory in potential. Within existing capitalist systems, the asymmetric relations between individual people and tech firms are such that piecemeal approaches, while obviously laudatory, will fail across and between the scales at which these firms operate. Uber, for example, is well known to purposefully ignore existing laws and regulations that would adversely affect its operations up until (and even after) it is legally challenged and forced to comply. The massive spending on California Proposition 22 in 2020, which effectively removed the right to minimum wage, sick leave, and other basic protections from gig-economy workers, demonstrates how regulation can and will be subverted by large, monied interests (see Chapter 4 for greater discussion).

While continuing to press for regulation and legislation, we suggest *also* turning towards other practices—practices that reshape how we engage with spatial data in our daily lives and simultaneously use said technologies to push towards wider-scale radical political change. The first step, we've argued, is to inform our daily device use with a radical praxis.

Geographer Greg Downey (2002) used the history and plight of telegraph messenger boys within the United States to illustrate the long history of labor exploitation that always undergirds technologies. More recently, the work of internet geographer Mark Graham and his colleagues has well illustrated how similar processes continue within the so-called gig economy (Graham et al. 2017). In light of this, in Chapter 1 we leveraged the works of late nineteenth- and early twentieth-century thinkers to demonstrate how new sociotechnical regimes are entwined with capitalist development and exploitation. Our goal in focusing on Heidegger, Debord, and Marcuse as "paths out" is not to suggest that they are the only notable scholars in this area, but instead to draw forward through the intervening decades a specific set of pertinent theoretical questions and critiques.

This informs our action. At its most fundamental, these ideas simply refuse to allow Google, Facebook, Amazon, Twitter, or any others to obfuscate spatial data exploitation. We can and do know, and *through that knowing* are able to make informed decisions on, how we do (and do not) make use of spatial data. This is the first call to action of this book: pause and be intentional in your use of digital technologies to the degree that you are able. As Chapter 3 outlined, "escape" is mostly a fiction offered to the elite, a conspicuous marker of (in)consumption; but, intentionality in use is open to all.

BACK TO A (DATA) FUTURE

The second call to action for our practices of informed use is for a (re) turn to détourned, digitally informed practices. Following from the work of Precarias a la Deriva's question, "What is your strike?", we must also ask ourselves (and our data selves) "*Where* is our strike?" Within the situated entanglements of a data-suffused world, where are the places and moments for potential resistance, for the building of solidarities? If we can intentionally understand our use of devices and our production of data through considerations of acceptance, resistance, making present, and escape (Chapter 3), then emergent strategies of resistance can begin to be formed via new acts of dérive, détournement, and ultimately, strikes (Chapter 4).

Such actions begin with the exploratory, such as the dérive. By intentionally opening for consideration the data that constitute ourselves within larger systems, we are able to unpack the moments, spaces, and even peoples that are elided in said systems. Whether this is something as banal as the path to a coffee shop or as profound as the oft-unmarked sites where large-scale computation occurs, the dérive unifies technology, politics, and self for consideration and contestation. Détournement takes data both from the drift and other sources and repurposes them, subverting the spectacle back upon itself in a way that calls attention to and inspires resistance. From creative hacking of corporate APIs and the creation of maps like Inside Airbnb to the reflexive, situated politics of groups like the Anti-Eviction Mapping Project, data détournement draws much from the practices of counter-data.[3] These efforts lead us back to the question, "Where is our strike?" Where Precarias a la Deriva found ways to mobilize without traditional forms of labor support, data strikes call us to do the same in the face of technical systems imbricated with many of our daily practices. If moments like #DeleteUber demonstrate the potential power carried in intentional use of technology, dérive, détournement, and strikes can help us realize them more.

ALREADY-EXISTING POLITICS AND TECHNO-UTOPIANISM

Even as digital technologies reshape society, a popular theoretical stance within the western critical tradition over the last 30 years has been to suggest that we live within a "post-political" world. In a simplified form, these arguments take Margaret Thatcher's[4] insistence of "no alternative" to heart: the end of the Cold War signaled a radical curtailing of what can be thought and what can be done as a broad politics of (neoliberal) consensus emerged. Post-political theorists argue that true politics exists only in moments of rupture from what already is, offering radical alternatives that, in a post-political frame, are always, if not already, inevitably co-opted.

To a degree, there is much value in such thinking. Occupy Wall Street was a moment of rupture that has become a slogan purchasable on t-shirts. But these ideas go too far in insisting on a purity test for what counts as political, one that relies upon a fundamentally antagonistic relation between what is and what might be. As geographer James McCarthy (2013) has quipped, "We Have Never Been 'Post-political.'" To suggest that current society and culture are non-political mistakes western state-scale power relations for a universal, hegemonic politics while simultaneously eliding the very real, daily practices that constitute politics across multiple scales and through multiple times.

Spatial data and the algorithms which analyze them clearly operate asymmetrically on users in ways which enable and constrain what can be known and done with and in the world. Cynical readings of data and their analyses can easily place them as yet another tool delimiting political possibilities, organized predominantly under the banner of multinational corporations and myriad apps in lieu of traditional state actors. However, they can *also* facilitate emergent means to subvert, control, and contest current regimes and relations. That is, after all, the point of this book—not simply to interpret, but to change our relations to spatial data and technologies. Such changes occur through daily practices and emergent strategies.

Geographic technologies can surely be used in moments of political rupture, such as when pro-democracy protestors in Hong Kong used mesh networks to avoid kettling efforts by security forces. Under such circumstances, certain technologies can serve profoundly liberatory roles by allowing for communication and coordination beyond the limits of participants' bodies and in the face of overwhelming use of force by the state. Nevertheless, as important as these uses are in specific moments, we cannot lose sight of the influences upon and uses of spatial data and technologies in everyday contexts. Within the banal moments of our lives, we can and must also put geospatial data to work in ways that move towards and further the causes of radical political change.

There is a certain brand of technological optimism, even solutionism, that looks to technological advances, and often the industry titans behind large tech firms, as potential saviors. Who will stop climate change? Elon Musk's solar-powered houses! How will we defeat COVID? Apple and Google have an app for that! Of course, as the failures of COVID-19 tracking apps across the world demonstrate, this brand of techno-utopianism is a form of magical thinking. An Apple Watch cannot accurately replace human contact-tracing efforts for COVID-19 because it was not materially designed to do so. The ways in which it represents the body, the ways that it tracks it, are ways designed to enable and encourage certain forms of consumptive practices, not to represent the whole self and the many physical, embodied entanglements of daily practice. Once more, the gap between the individual that capital can see and the individual's *being-in-the-world* presents an insurmountable problem for technology designed to produce profit.

Rather than turn towards the magic of technologies and the billionaire titans that currently steer them, we must turn to one another, to our shared humanity, and to the politics of the everyday. We live in the ruins of a world wrought by the global pursuit of capital accumulation, a world still beholden to the extraction and consumption of fossil fuels even as every indicator suggests this will cause social and ecological devastation at heretofore unseen scales. Climate change, white supremacy, toxic masculinity, and infectious diseases push at the limits of neoliberal concepts of personal responsibility as, over and over again, fantasies of individual control fall in the face of such systemic crises. But that doesn't mean there are no alternatives. Rather, intentional individual actions, as we note above, must link together to form something larger; an emergent politics that builds different kinds of data regimes from the ground up. We must find our strikes and exercise them.

WE[5] LIVE IN THE GAPS

It may seem counterintuitive to insist on both the asymmetrical relationship between individuals and large technology firms *as well as* the need to foster intentional use and therein emergent technological solidarities. Threading that needle is not only possible, but necessary. Without losing sight of how data tools are currently designed to predict, exploit, and delimit daily life, we must also recognize the profound opportunities they can open for us.

The processes of capitalism require an ever-expanding market, and that currently includes corkscrewing inwards, ever-increasing marketization of our thoughts, our attention, and our bodies. This process has an expansionist, totalizing tendency; it is not a static, complete totality. New technologies suffuse outwards across space and inwards into bodies and daily movements, attempting to capture ever more personal time and energy, attempting to guide and shape daily experiences. Nevertheless, due to its nature as a dynamic, imperfect process, there are still gaps and cracks. There are still moments of privacy, moments of personal serendipity. Yelp may guide us to the nearest Thai restaurant and recommend a certain dish, but it does not yet control how we greet a friend we meet there nor how much we tip our server. Perhaps it will nudge the latter, and certainly apps have emerged that attempt to do so (Uber Eats, Instacart), but even as these technologies attempt to establish norms, some gaps remain and new ones open. What's counted counts, but that is not and can never be all that is. A mesh network to avoid kettling, a counter-map that facilitates affordable housing—technology can be détourned, it can be repurposed. We live in those cracks, in those gaps between what is and what might be.

This builds to our third call, to find and live in the cracks, to widen them, through acts of intentionality, repurposing, and resistance. First, we must be intentional in our use of digital technologies. In part, this means conscious decision-making in how we do and do not use devices, an understanding of the stakes and a knowing attempt to delimit our own exploitation. Intentionality also means refusing to allow just-so stories of techno-utopianism to efface the underlying motivations and very real histories of failure. "Who could have known?" has been used far too long to abdicate from any responsibility by technology firms even as, in the same breath, they promise a new suite of solutions to problems of their own creation. "Download our ad-blocker!" Second, our intentionality should be aimed towards the active deconstruction and reconstruction of spatial data in our lives. We must find our strikes, whether they be deleting an app to highlight Uber's complicity in strike breaking (Siddiqui 2017) or building coordination and mutual aid among precarious users and gig workers. We must explore and better know our own data to make them ours, to repurpose them in ways that build solidarities with others. Greater regulation may help, but the methods of dérive, détournement, and strikes are directly useful, even as they are not intended to be the only techniques.

Putting the first two calls into practice helps produce the spaces and places of the third. Living in the gaps is not a total escape. It is situated, always incomplete and temporary, and yet it does provide benefits. It facilitates greater agency, not only over data, but through those data to a broader being in the world. After all, the value of going places to relax on a day off is not just the production of a location history; it is the experience of being there. It can also allow for greater solidarity, rather than the constant gamesmanship of marketized relationships, particularly in personal relationships of the everyday.

Living in the gaps is not a call for everyone to learn how to code as some sort of panacea for public good. Being able to program is only one of many means of living in the cracks. While it can be useful to learn to write code, far too often it is forced into curriculums and careers as some modern-day tonic to cure the ills of rampant data capitalism. In the politics of resistance to spatial data, there is no room for a techno-elitist vanguard. Rather, there are a panoply of situated means by which to act with, through, and against spatial data and the algorithms which sort daily life.

One example is part of this book. We developed and will continue to maintain *DataResistance* (https://github.com/DataResistance) as a collaborative public repository. It is a space for both ourselves and readers to share not just code and computational tools, but also techniques and stories of life in the cracks. From the ability to create maps of public running routes in a city to guides for personal data dérives (both available at time of writing), the repository is not an arcane location for cryptic Python code, but a living, breathing document of our struggles. While we encourage every reader to consider participating, that's hardly necessary. It is but one space of consideration, one example of a praxis that must run through all of life.

OUR DATA ARE US, SO MAKE THEM OURS

Multiple subcultures indicate a growing interest in making the data that surrounds us tangible and visceral. One craft-based approach is knitting temperature blankets, blankets in which the color of each row of knitted yarn represents the temperature at a geographic location over a time period. Table 5.1 specifies a range of colors for a temperature blanket for Tacoma, Washington, USA.[6] Each color corresponds with a temperature range. The blanket emerges by knitting one or two rows of yarn each day for a year in the color determined by the closest weather station. After a year, or other period of time, the blanket reflects the shifting patterns of weather in the region, with each row of color recording a day of temperature data. This is but one example of a larger trend in which individuals and groups use physical crafting to represent, visualize, and interpret data produced in their lives. Alice Thudt, Uta Hinrichs, and Sheelagh Carpendale have termed this practice "data craft" and see it as "a way to create meaningful physical mementos based on digital records of personal and shared experiences" (Thudt et al. 2017, 2). It is meant to both integrate and make present digital data in everyday life and to inspire reflection upon the creation and interpretation of these data.

Table 5.1   Suggested yarn colors for a temperature blanket based on Tacoma, WA

Temperature (°F)	Color
>101	Crimson
91–100	Raspberry
81–90	Tropical Pink
71–80	Radiant Yellow
61–70	Mint
51–60	Forest
41–50	Sea Blue
33–40	Orchid
<32	Mixed Berry

These sorts of projects embody a personal intentionality and reflective practice around digital data, but they also highlight the selective nature of data. *Why this color for that temperature? Why track calls to one's parents instead of one's lover(s)? How many days to record?* In the case of a temperature blanket, these are all personal decisions made by a friend of the authors who knits, but they also reflect the limits and importance of quantification. In these projects, the crafters choose, within the limits of the technology, what to record and how to visualize and interpret it— in short, how to carry it with them in daily life. In regimes of spatial data, those decisions are predominantly made by large, profit-seeking corporations in ways that reduce lived experience into a stream of exchangeable commodities. *Data on individuals that like pro wrestling and kittens are worth $1 per 1,000 email contacts, whereas data on individuals who like monster truck rallies and opera are worth $2.*

In both the case of the blankets and the tech companies, the data are individualized. But only in the crafter's case does the individual choose what data and how they are represented. The spectacular individual that spatial data constructs is an individual made of the data produced by and for the generation of profit. However, they are produced through much of the same sorts of data that data crafters use to make far more personal, reflexive objects. Data crafting projects, data hacking projects, data dérives, and other opportunities to explore and take control of the data produced through our digital devices hold promise for improving our lives and making society more equitable. Spatial data technologies are impoverished by a one-dimensional pursuit of profit. They have much to offer, for they allow us to reach beyond the limits of our senses to experience and come into contact with new knowledges, places, peoples, and ways of living. The data we produce are us, and the future demands that we make them ours.

Epilogue

The precise algorithms, apps, and data formats through which we engage the world are constantly evolving, as noted throughout this book. A tactic that works today may be co-opted into a value-producing act of data dispossession in the future; an app or API may disappear or change its terms of service in new and unexpected ways. Even now, despite claims of the end of moving fast and breaking stuff, venture capitalists continue to bet on the ability of any given startup to successfully disrupt and envelop existing modes of living and social relations.

This book was conceived of during the summer and fall of 2019, and written predominantly during the long pandemic-inflected months of 2020 between periods of no childcare. As such, it necessarily focuses on moments and examples that came before that time; but the fundamental relations between spatial data, individual, and society remain. Apple's new (at time of writing) iOS 14.5 helps to illustrate this point. iOS 14.5 introduces a new feature that requires all applications to ask for explicit permission from the user in order to track any data between applications or websites. In response, Facebook and Instagram have begun to warn users that only through allowing tracking can they "help keep [the services] free" (Haslam 2021).

While this beautifully illustrates what "free" means in the context of the social media giant, it's also rather beside the point with respect to larger structures of data dispossession. First, there's the obvious tension between Apple's attempts to keep users and their data within Apple's monopolistic ecosystem and Facebook's requirement to extract that self-same data for its own profits. It's not that Apple devices aren't generating huge swathes of data, it's that they aren't letting their competitors access that data without your consent. Second, even as Apple nominally steps up its transparency with respect to data privacy, new services like Amazon Sidewalk come online to offer exciting new moments of data dispossession.[1] The technologies change, the capitalist imperatives don't.

This is why, again and again, we cannot allow the terms of our engagement with technology to be solely dictated by profit motives. In *Dear*

Science, professor of Gender Studies and pillar of black geographies Katherine McKittrick outlines the importance in no uncertain terms:

> [Algorithms] are anticipatory computations that tell us what we already know, but in the future. If we want different or better or more just futures and worlds, it is important to notice what kind of knowledge networks are already predicting our futures.
>
> (McKittrick 2021, 116)[2]

While we have not focused explicitly on race, the tendency of current algorithms and data to simultaneously perpetuate and elide exploitation and injustices bears repeating. Better, more equitable alternatives are not built-in or guaranteed. If we allow *what is* to determine *what might be*, we are simply reinforcing an existence defined by the ability to consume and predicated upon denying humanity to those who do not conform to that white, patriarchal, wealthy norm. We must do better. The priorities that drive the design and structure of technology and data are social choices. They could reflect other social imperatives and fulfill users' needs differently. The use of technologies, while more limited, also allows for limited considered choice. And through resistance and consideration, not just in theory or isolated moments, but through everyday life, we can do better.

Notes

INTRODUCTION

1. In this book, we consider data as a generic concept to be plural, but specific ideological instantiations—such as "big data"—to be singular. Grammatically, we have followed professor Anna Lauren Hoffman's suggestion of replacing every instance of data with "giraffes" to aid in number agreement. Any remaining mistakes are our own.
2. This interpretation of the Frankfurt School is not universally accepted even with respect to its use with geospatial technologies (see, for example, Eades' (2010) response to Kingsbury and Jones). We return to and develop this complicated intersection of ideas in the first chapter.
3. For more on this case, see Chapter 2.
4. XML itself is an application of the Standardized Generalized Markup Language (SGML), developed in the 1980s, which is, in turn, derived from IBM's Generalized Markup Language (GML) first created in the 1960s, the point here not being some pedantic, arcane walk through the histories of markup languages for digital documents, but to once more stress that the roots of a given technology are often much deeper and far more entangled than popular narratives might suggest.

1 LIFE IN THE AGE OF BIG DATA

1. Parts of this chapter may be found in Thatcher's 2014 dissertation "Mobile Navigation Applications: Hidden Ontologies, Epistemic Limits, and Technological Teleology," but the contents here have been heavily expanded and altered for accessibility by a general audience.
2. In his work, a new Satanic Mill can be found in the "computerized prison" which promises "prison[s] managed like an orderly factory" in which subjects are driven from point to point within the system to produce a predictable, orderly system of "great efficiency and scale" (Jefferson 2020, 78). Here, we extend this analogy in line with the ways in which data capitalism attempts to structure, predict, and control everyday life through processes of data dispossession. Any misapplication of his precise and excellent point is our own.
3. For an excellent discussion on Technological Determinism and Karl Marx, see Bimber (1990). After surveying how others have written on Marx's views, Bimber concludes that Marx's view of technology does not make an ontological claim on the universal laws of nature and therefore is not truly

deterministic. Sayer's *Capitalism and Modernity: An Excursus on Marx and Weber* (1991) also explored this area.

4. See also Weber (2008 [1891]).

5. This view puts him at odds with some of the writings of Engels and, much later, Althusser (see, for example, Althusser 1962). This concept will become far more important as we begin to discuss the emergent logics of algorithmic approaches to big data.

6. In the same paragraph of *The Protestant Ethic and the Spirit of Capitalism* as the previous quote (Weber 2005[1930], 123), Weber wonders whether the world and those who live within it might remain so trapped "until the last ton of fossilized coal is burn[ed]." In his own oblique way, Weber nods at the second contradiction of capitalism as environmentally unsustainable as a passing note (see O'Connor 1998).

7. While we retain the original language's gender within citations, we use "they" as a gender-neutral third-person singular and plural pronoun in our own text.

8. Although outside the purpose of this work, readers familiar with the Frankfurt School might note that a major point of contention between Benjamin and Theodor Adorno lay in their disagreement over the importance of popular versus avant-garde art as revolutionary tools. Adorno insisted that Benjamin overstated the significance of the proletariat as producers of works of significance. Letters between the two directly addressing this can be found in Adorno et al. (2007 [1977]).

9. This can be found perhaps most clearly in Marx's discussion of the dual character of labor embodied in linen and coats in Chapter 1 of *Capital Volume 1* (Marx 1990 [1848], 131).

10. The key theoretical movement here is to see the "culture industry" as equivalent to a factory in terms of what it produces and how it organizes labor. However, instead of widgets, it produces an increasingly standardized set of cultural goods such as movies, magazines, radio and television shows, and so on, producing, as Bruce Springsteen notes, a situation in which we have 57 channels, but nothin' is on.

11. A more general lament for the reduction of science to neoliberal counting, and a plea and pathway to alternatives, can be found in Isabelle Stengers' *Another Science Is Possible: A Manifesto for Slow Science* (2018).

12. Although the exact number and nature of the contradictions of capitalism are debated (see, inter alia, O'Connor 1991; Harvey 2015), the "first" one identified by Marx refers to the fact that as capital exercises power over labor in order to increase profits (for example, by lowering wages or reducing the number of employees), workers are increasingly unable to purchase those very goods.

13. Listing these separately is in no way meant to imply they are not interconnected, entwined braids of the same processes.

14. See Mitchell and Trawny (2017) for an engaged discussion on the uses and meanings of Heidegger's philosophy in light of its anti-semitism.

15. It is a foundational part of Feenberg's thought on technology that we covered earlier in this chapter.

16. Writing for *MIT Technology Review*, Karen Hao (2020) has an excellent summary of the matter.

17. Picking apart this quote has become something of a shibboleth amongst those writing critically on data and technology, yet—at the same time—its underlying ideology remains popular amongst mainstream data scientists. For more discussion on the role this quote has played in critical discourse, see the introduction of Thatcher et al.'s *Thinking Big Data in Geography* (2018).

18. In *The Ticklish Subject*, Žižek (1999, 62) provides a surprisingly succinct definition of *being-in-the-world* as an individual's existence within a "concrete and ultimately contingent life-world."

19. In Chapter 4, we develop the techniques of data dérive and data détournement, suggesting their importance in practices that build shared solidarities with and through the data spectacle.

20. And that of his students (Feenberg 1999).

21. Ruser later corrected that it was based upon 3 billion GPS points: https://twitter.com/Nrg8000/status/957488086144892928 (last accessed July 2021).

22. https://twitter.com/nrg8000/status/957318498102865920 (last accessed July 2021).

23. For non-British readers, the Home Office is the government agency/ministry charged with immigration enforcement within the United Kingdom's borders, akin to Immigration and Customs Enforcement (ICE) in the United States.

24. Data on non-EU nationals were not published.

25. We introduced this term in "Data Colonialism through Accumulation by Dispossession" (Thatcher et al. 2016), drawing on Morgan Robertson's "The Nature that Capital Can See" (2004).

26. An API is a standardized means by which one software program can incorporate data or services from another. For example, weather forecasting websites often include an embedded Google Map (provided through the Google Maps API) with the forecaster's weather data plotted within Google's map. As the name suggests, it provides an *interface* through which code from different sources can communicate. An API specifies both what can be asked and the format in which the answer will come. It is a means of abstraction through which an individual program doesn't need to know *how* an analysis is conducted, just what the results are. Common analogies include the menu in a restaurant (an order usually does not specify all of the ingredients or how they are cooked) or managing a bank (a customer often does not know the password to the vault, but can access their money from it).

27. The full report has been referred to as "The Drone Papers" and is available at https://theintercept.com/drone-papers/ (last accessed July 2021).

28. In an interesting twist, the change to the Foursquare API affected other Foursquare-based applications. For example, Assisted Serendipity, an early application that simply displayed the ratio of male-to-female check-ins at

nearby establishments was also disabled by this change. Assisted Serendipity had previously been specifically mentioned by Foursquare's Chief Executive Officer (CEO) as precisely the type of application they wished to see built using their API. See Thatcher (2014) for a more intensive look at the implications of this process, but don't feel too badly for the developers of Assisted Serendipity, as their next start-up was purchased by Airbnb.

2 WHAT ARE OUR DATA, AND WHAT ARE THEY WORTH?

1. This does illustrate the ineffective nature of arguments for corporations paying individuals for their data, so-called "data dividends."
2. Grand View Research, a market intelligence firm, has placed the geospatial analytics market at north of $50 billion per annum in 2018, and growing rapidly (Grand View Research 2018).
3. We say "people," not just "users," because many companies, from Facebook to Palantir, extract and analyze data even from those who do not use their services.
4. In a piece in *The Atlantic*, Ingrid Burrington (2016b) similarly observed: "Networks build atop networks."
5. In a 2015 study, Durairajan et al. found that co-occurrence of fiber-optic cable with road or rail infrastructure was high throughout the network, but highest was co-occurrence with both road and rail. In other words, where highways and rail lines converge, so too does fiber-optic cable.
6. In an interesting, informed piece, Chuncheng Liu (2019), a PhD candidate in Sociology and Science and Technology Studies at University of California San Diego, argues that the social credit systems of China are best understood as symbolic systems of performative power that are best contextualized within the political and social histories of the People's Republic of China.
7. Nebulous "terrorist no-fly lists" exist in many western countries as well, often with the exact criteria for being placed upon or removed from them shrouded in state secrecy and bureaucracy.
8. *Carpenter v. United States* (2018) and *United States v. Jones* (2012).
9. One of the more insidious ideological moves by the Chinese state was reframing its surveillance for COVID-19 in terms of a war against the disease. A Reuters report by Cate Cadell (2020), features quotes emphasizing a "war situation" and the need for "war-time thinking." In the US, "war on terror" can stand in as a "black mirror," making clear how such frames allow for the normalization of increased surveillance with an enemy that is both invisible and unvanquishable.

3 EXISTING EVERYDAY RESISTANCES

1. For collective actions, see Chapter 4.
2. These three examples follow the tale of British Airways' 2018 data breach. First, the breathless explanations of "how," then the regulatory authorities' action, and finally the consumer class action lawsuit. Rinse, repeat.

3. Brook Gladstone's 2013 *On the Media* interview with George Washington University law professor Daniel Solove on Erich Schmidt's quote is an excellent discussion of the dangers of reductive, teleological thinking regarding privacy: www.wnycstudios.org/podcasts/otm/segments/260644-if-youve-got-nothing-hide-youve-got-nothing-fear (last accessed July 2021).

4. Even without a "smart" phone, any cellular device produces temporal geospatial data, the simple act of connecting to a provider's network records the phone's ID, the closest tower, and the time of connection. Tied together, these "dumb" data (as in not generated through any "smart" features) provide startlingly granular time-travel patterns for individual (or at least their phones) and constitute billions of dollars in data sales annually for network providers.

5. Defunct (and impossible) now, the Institute for Applied Autonomy created iSee, a tool for avoiding surveillance cameras, in the early 2000s. This interactive web map provided directions meant to avoid the 2,400 surveillance cameras the group identified in New York City. Even then, as Erik Baar noted in *Wired* in 2001, "[a] 12-block walk down Park Avenue becomes a 35-block trek when you avoid the surveillance cameras."

6. Another noted flip phone user, Warren Buffet, upgraded to an iPhone in 2020 (Bursztynsky 2020).

7. 911 is the emergency number in the United States.

4 CONTESTING THE DATA SPECTACLE

1. At release, one of the cited examples of the need for an API was a map of tweets called "This world is small!" The original announcement can be found here: https://blog.twitter.com/official/en_us/a/2006/introducing-the-twitter-api.html (last accessed July 2021).

2. A complete description of the technical specifications of DOLLY may be found here: http://www.floatingsheep.org/p/dolly.html (last accessed July 2021).

3. RIP Hana Kimura.

4. In a slightly bemusing bit of corporate speak, in its 2019 fiscal year report to the Securities and Exchange Commission (Twitter 2019, 5), Twitter eschews the term "user" in order to show "empathy" for those that make use of its platform. We apologize for our lack thereof.

5. What data are stored, how they are communicated, and so on all have epistemological effects—see Thatcher (2014) for more discussion.

6. Without belaboring the point, the "base" for Marx would be the relations of material production, a factory making cars or the like, while the "superstructure" would be, roughly, everything else, the realm of ideas and culture. For Marx, the base *determined* the superstructure (for more, see Williams 1973).

7. Then a record for spending on a California voter proposition.

8. The '20 minute neighborhood' is a neoliberal fantasy of urban planning in which equity and sustainability are achieved by living in an area in which most (though not all) needs can be met through a 20 minute walk, bike, or public transit ride. Perhaps most famous for its implementation in Melbourne, Victoria, Australia (Stanley and Hansen 2020), it has also gained traction in other cities such as Portland, Oregon, USA and even been endorsed by the American Association of Retired Persons (Walljasper 2017).

9. We're geographers. Spatial data are our stock in trade. A successful non-geographic data drift may be possible: If you see a way, go for it!

10. At time of writing, the film is readily available on many streaming sites, such as YouTube.

11. Though Banksy has his own particular critique of art and commodification.

12. Sousveillance was first defined by Mann (1998), and refers to the processes by which individuals may turn normally surveillant technologies back upon those in power—for example, filming and uploading police officers. More recently, works like Kitchin (2014) have broadened the term to refer generally to data collected by individuals.

13. A Markov chain refers to a mathematical system that models all of the potential states of the system and the probability of moving from one state to another. As Powell and Lehe (2014) explain, a Markov chain model of a baby's behavior might include "playing," "eating," "sleeping," and "crying" with the model telling you the probability of moving from one state (eating) to another (sleeping). In the case of @Marxbot1, the model records the probability for one word to follow another within a given corpus.

14. @Marxbot1 is run by Jim Thatcher, the code is available in the book's online repository of code at https://github.com/DataResistance. We particularly encourage users to combine works by various authors, such as Ursula Le Guin and Guy Debord, to produce new creations.

15. Links to a variety of police violence and other mapping projects can be found at the book's GitHub repository (and living appendix), *DataResistance* (https://github.com/DataResistance).

16. Kettling refers to the practice by police forces of confining protestors into a small, controlled area, often to arrest them. Ostensibly for crowd control, in practice such techniques often result in tensions boiling over (like a tea kettle on a flame), providing police forces the legal justification for violence.

17. Facing intense backlash, Apple restored the HKmap.live app shortly after its removal, and eventually issued a corporate statement on human rights in August of 2020.

5 OUR DATA ARE US, SO MAKE THEM OURS

1. The A-level exams are a set of college qualification exams offered in England. They are not mandatory, but are a major factor in the university admission process.

2. Sarah Hunt's 2014 piece on the ontologies of indigeneity offers a powerful examination of the "ongoing (neo)colonial relations that shape geographic knowledge production" and their limits (Hunt 2014, 27).

3. Our purposes here are not to suggest that these terms define the projects listed, but rather to give examples of exemplary projects that, in ways, align with the core techniques.

4. Thankfully, no relation.

5. Any use of "we" should raise the immediate question of who is included in such a construction and, consequently, who is or might be excluded. Here, our intent is a generalized call to action for those who find themselves within the data spectacle.

6. Yarn colors provided by archaeologist and knitter Madelynn von Baeyer.

EPILOGUE

1. Installed on all Echo devices produced since 2018 and in many Ring surveillance and doorbell products, Sidewalk is a bandwidth-sharing technology that creates a mesh network allowing for other devices to access the internet through your internet connection, unless you opted out before the June 2021 deadline.

2. In her book, McKittrick eloquently pushes against extractive academic citational practices; here, we have included a portion of her much larger thoughts as an attempt to think with while acknowledging; a necessarily imperfect practice within an academic discourse that still prizes citation counts and impact factors. For an in-depth discussion of citational practice and its impacts, we recommend Carrie Mott and Daniel Cockayne's excellent article "Citation Matters: Mobilizing the Politics of Citation toward a Practice of 'Conscientious Engagement'" (2017).

Bibliography

ACLU of Illinois. (2011) *Chicago's Video Surveillance Cameras: A Pervasive and Unregulated Threat to Our Privacy.* www.aclu-il.org/sites/default/files/wp-content/uploads/2012/06/Surveillance-Camera-Report1.pdf (last accessed June 2021).

Adams, Richard, and Stewart, Heather. (2020) "Boris Johnson Urged to Intervene as Exam Results Anger Escalates." *The Guardian.* www.theguardian.com/education/2020/aug/16/boris-johnson-urged-to-intervene-as-exam-results-crisis-grows (last accessed January 2021).

Adorno, Theodor, Benjamin, Walter, Bloch, Ernst, Brecht, Bertolt, and Lukacs, Georg. (2007[1977]) *Aesthetics and Politics.* London: Verso Books.

Aldhous, Peter. (2016) "The FBI Used Its Most Advanced Spy Plane to Watch Black Lives Matter Protests." *Buzzfeed News.* www.buzzfeednews.com/article/peteraldhous/fbi-surveillance-plane-black-lives-matter-dc (last accessed February 2021).

Aldhous, Peter, and Seife, Charles. (2016) "Spies in the Skies." *Buzzfeed News.* www.buzzfeednews.com/article/peteraldhous/spies-in-the-skies (last accessed January 2021).

Althusser, Louis. (1962) "Contradiction and Overdetermination." Translated by Ben Brewster. www.marxists.org/reference/archive/althusser/1962/overdetermination.htm (last accessed November 2020).

Alvarez León, Luis F. (2018) "Counter-mapping the Spaces of Autonomous Driving." *Cartographic Perspectives* 92: 10–23.

Amoore, Louise. (2020a) *Cloud Ethics: Algorithms and the Attributes of Ourselves and Others.* Durham, NC: Duke University Press.

Amoore, Louise. (2020b) "Why 'Ditch the Algorithm' Is the Future of Political Protest." *The Guardian.* www.theguardian.com/commentisfree/2020/aug/19/ditch-the-algorithm-generation-students-a-levels-politics (last accessed January 2021).

Anderson, Chris. (2008) "The End of Theory: The Data Deluge Makes the Scientific Method Obsolete." *WIRED.* www.wired.com/science/discoveries/magazine/16-07/pb_theory (last accessed December 2020).

Andrejevic, Mark. (2014) "The Big Data Divide." *International Journal of Communication* 8: 1,673–1,689.

Archibong, Ime. (2018) "An Update on Our App Investigation and Audit." *Facebook.* https://about.fb.com/news/2018/05/update-on-app-audit/ (last accessed January 2021).

Aronson, Ronald. (2014) "Marcuse Today." *Boston Review.* https://bostonreview.net/books-ideas/ronald-aronson-herbert-marcuse-one-dimensional-man-today (last accessed December 2020).

Arrieta-Ibarra, Imanol, Goff, Leonard, Jiménez-Hernández, Diego, Lanier, Jaron, and Weyl, E. Glen. (2018) "Should We Treat Data as Labor? Moving beyond 'Free.'" *Papers and Proceedings of the One Hundred Thirtieth Annual Meeting of the American Economic Association* 108: 38–42.

Baard, Erik. (2001) "Routes of Least Surveillance." *WIRED*. www.wired. com/2001/11/routes-of-least-surveillance/ (last accessed January 2021).

Baltrusaitis, Justinas. (2019) "Top 10 Countries and Cities by Numbers of CCTV Cameras." *PreciseSecurity*. www.precisesecurity.com/articles/Top-10-Countries-by-Number-of-CCTV-Cameras (last accessed January 2021).

BBC News. (2018) "Fitness App Strava Lights Up Staff at Military Bases." www. bbc.com/news/technology-42853072 (last accessed December 2020).

BBC News. (2020) "Elon Musk Says Full Self-driving Tesla Tech 'Very Close.'" www.bbc.com/news/technology-53349313 (last accessed January 2021).

Beller, Jonathan. (2012) "Wagers within the Image: Rise of Visuality, Transformation of Labour, Aesthetic Regimes." *Culture Machine* 13. https:// culturemachine.net/wp-content/uploads/2019/01/466-977-1-PB.pdf (last accessed January 2021).

Benjamin, Ruha (ed.). (2019a) *Captivating Technology: Race, Carceral Technoscience, and Liberatory Imagination in Everyday Life*. Durham, NC: Duke University Press.

Benjamin, Ruha. (2019b) *Race after Technology: Abolitionist Tools for the New Jim Code*. Cambridge, UK: Polity Press.

Benjamin, Walter. (1978) *Reflections: Essays, Aphorisms, Autobiographical Writings*. Translated by Peter Demetz. New York: Schocken Books.

Benjamin, Walter. (2008) *The Work of Art in the Age of Its Technological Reproducibility and Other Writings on Media*. Jennings, Michael W., Doherty, Brigid, and Levin, Thomas Y. (eds.). Cambridge, MA: The Belknap Press of Harvard University Press.

Benjamin, Walter. (2008[1935a]) "Paris, the Capital of the Nineteenth Century." Translated by Howard Eiland. In: Jennings, Michael W., Doherty, Brigid, and Levin, Thomas Y. (eds.) *The Work of Art in the Age of Its Technological Reproducibility and Other Writings on Media*. 96–115. Cambridge, MA: The Belknap Press of Harvard University Press.

Benjamin, Walter. (2008[1935b]) "The Formula in Which the Dialectical Structure of Film Finds Expression." Translated by Edmund Jephcott. In: Jennings, Michael W., Doherty, Brigid, and Levin, Thomas Y. (eds.) *The Work of Art in the Age of Its Technological Reproducibility and Other Writings on Media*. 340–341. Cambridge, MA: The Belknap Press of Harvard University Press.

Benjamin, Walter. (2008[1936]) "The Work of Art in the Age of Its Technological Reproducibility: Second Version." Translated by Edmund Jephcott and Harry Zohn. In: Jennings, Michael W., Doherty, Brigid, and Levin, Thomas Y. (eds.) *The Work of Art in the Age of Its Technological Reproducibility and Other Writings on Media*. 19–55. Cambridge, MA: The Belknap Press of Harvard University Press.

Benjamin, Walter. (2008[1938]) "The Telephone." Translated by Howard Eiland. In: Jennings, Michael W., Doherty, Brigid, and Levin, Thomas Y. (eds.) *The Work of Art in the Age of Its Technological Reproducibility and Other Writings on Media*. 77–78. Cambridge, MA: The Belknap Press of Harvard University Press.

Bennett, Shea. (2014) "Twitter Buys Gnip, Promises to Make Data More Accessible." *AdWeek*. www.adweek.com/performance-marketing/twitter-buys-gnip/ (last accessed August 2021).

Berry, David M. (2011) *The Philosophy of Software: Code and Mediation in the Digital Age*. London: Palgrave Macmillan.

Bilton, Nick. (2012) "Girls Around Me: An App Takes Creepy to a New Level." *Bits, The New York Times*. https://bits.blogs.nytimes.com/2012/03/30/girls-around-me-ios-app-takes-creepy-to-a-new-level/ (last accessed November 2020).

Bimber, Bruce. (1990) "Karl Marx and the Three Faces of Technological Determinism." *Social Studies of Science* 20(2): 333–351.

Bishop, Jordan. (2018) "This Is How 380,000 British Airways Passengers Got Hacked." *Forbes*. www.forbes.com/sites/bishopjordan/2018/09/11/how-british-airways-got-hacked/ (last accessed January 2021).

Boellstorff, Tom. (2013) "Making Big Data, in Theory." *First Monday* 18(10). https://firstmonday.org/ojs/index.php/fm/article/view/4869/3750 (last accessed January 2021).

Borach, Anne. (2007) "Cell Carriers Fined over Missed e911 Deadlines." *CNet*. www.cnet.com/news/cell-carriers-fined-over-missed-e911-deadlines/ (last accessed January 2021).

Bridle, James. (2018) *New Dark Age: Technology and the End of the Future*. London: Verso Books.

Brownlee, John. (2012) "This Creepy App Isn't Just Stalking Women without Their Knowledge, It's A Wake-up Call about Facebook Privacy." *Cult of Mac*. www.cultofmac.com/157641/this-creepy-app-isnt-just-stalking-women-without-their-knowledge-its-a-wake-up-call-about-facebook-privacy/ (last accessed December 2020).

Bryant, Martin. (2015) "Twitter to Cut off Firehouse Resellers as It Brings Data Access Fully In-house." *The Next Web*. https://thenextweb.com/dd/2015/04/11/twitter-cuts-off-firehose-resellers-as-it-brings-data-access-fully-in-house/ (last accessed January 2021).

Burn-Murdoch, John. (2012) "Church vs Beer: Using Twitter to Map Regional Differences in US Culture." *The Guardian*. www.theguardian.com/news/datablog/2012/jul/04/us-fourth-july-twitter-beer-church (last accessed January 2021).

Burrington, Ingrid. (2016a) *Networks of New York: An Illustrated Field Guide to Urban Internet Infrastructure*. New York: Melville House.

Burrington, Ingrid. (2016b) "Why Amazon's Data Centers Are Hidden in Spy Country." *The Atlantic*. www.theatlantic.com/technology/archive/2016/01/amazon-web-services-data-center/423147/ (last accessed January 2021).

Bursztynski, Jessica. (2020) "Billionaire Apple Investor Warren Buffet Finally Trades in His $20 Phone for an iPhone." *CNBC.* www.cnbc.com/2020/02/24/apple-investor-warren-buffett-traded-in-his-flip-phone-for-an-iphone.html (last accessed January 2021).

Cadell, Cate. (2020) "China's Coronavirus Campaign Offers Glimpse into Surveillance System." *Reuters.* www.reuters.com/article/us-health-coronavirus-china-surveillance-idUSKBN2320LZ (last accessed January 2021).

Calzati, Stefano. (2020) "Decolonizing 'Data Colonialism': Propositions for Investigating the Realpolitik of Today's Networked Ecology." *Television & New Media.* DOI: 10.1177/1527476420957267.

Carreyrou, John. (2018) *Bad Blood: Secrets and Lies in a Silicon Valley Startup.* New York: Knopf.

Casas-Cortés, Maribel. (2014) "A Genealogy of Precarity: A Toolbox for Rearticulating Fragmented Social Realities in and out of the Workplace." *Rethinking Marxism* 26(2): 206–226.

Casas-Cortés, Maribel, and Cobarrubias, Sebastián. (2007) "Drifting through the Knowledge Machine." In: Shukaitis, Stevphen, and Graeber, David (eds.) *Constituent Imagination: Militant Investigations, Collective Theorization.* 112–126. Chico, CA: AK Press.

Chacon, Benjamin. (2021) "The Best Time to Post on Instagram in 2021, According to 12 Million Posts." *Later.* https://later.com/blog/best-time-to-post-on-instagram/ (last accessed January 2021).

Chen, Brian X. (2018) "How to Protect Yourself (and Your Friends) on Facebook." *The New York Times.* www.nytimes.com/2018/03/19/technology/personaltech/protect-yourself-on-facebook.html (last accessed January 2021).

Chtcheglov, Ivan. (1953) "Formulary for a New Urbanism." http://library.nothingness.org/articles/SI/en/display/1 (last accessed January 2021).

Chu, Cecilia L., and Sanyal, Romola. (2015) "Spectacular Cities of Our Time." *Geoforum* 65: 399–402.

Clark, Liat. (2015) "Robots Are Increasing Our Wages, Not Stealing Our Jobs." *WIRED.* www.wired.co.uk/article/robots-are-not-stealing-our-jobs (last accessed November 2019).

Confessore, Nicholas. (2018) "Cambridge Analytica and Facebook: The Scandal and the Fallout So Far." *The New York Times.* www.nytimes.com/2018/04/04/us/politics/cambridge-analytica-scandal-fallout.html (last accessed January 2021).

Confessore, Nicholas, LaForgia, Michael, and Dance, Gabriel J.X. (2018) "Facebook Failed to Police How Its Partners Handled User Data." *The New York Times.* www.nytimes.com/2018/11/12/technology/facebook-data-privacy-users.html (last accessed January 2021).

Cook, Gary, Lee, Jude, Tsai, Tamina, Kong, Ada, Deans, John, Johnson, Brian, and Jardim, Elizabeth. (2016) *Clicking Clean: Who Is Winning the Race to Build a Green Internet?* Greenpeace. www.clickclean.org/downloads/ClickClean2016%20HiRes.pdf (last accessed July 2021).

Couldry, Nick, and Mejias, Ulises A. (2019a) "Data Colonialism: Rethinking Big Data's Relation to the Contemporary Subject." *Television & New Media* 20(4): 336–349.

Couldry, Nick, and Mejias, Ulises A. (2019b) *The Costs of Connection: How Data Are Colonizing Human Life and Appropriating It for Capitalism.* Oxford, UK: Oxford University Press.

Counter Cartographies Collective, Dalton, Craig, and Mason-Deese, Liz. (2012) "Counter (Mapping) Actions: Mapping as Militant Research." *ACME: An International Journal for Critical Geographies* 11(3): 439–466.

Cox, Kate. (2021) "Military Intelligence Buys Location Data Instead of Getting Warrants, Memo Shows." *Ars Technica.* https://arstechnica.com/tech-policy/2021/01/military-intelligence-buys-location-data-instead-of-getting-warrants-memo-shows/ (last accessed January 2021).

Crampton, Jeremy W., Roberts, Susan M., and Poorthuis, Ate. (2014) "The New Political Economy of Geographical Intelligence." *Annals of the Association of American Geographers* 104(1): 196–214.

Dahlqvist, Fredrik, and Patel, Mark. (2019) "Growing Opportunities in the Internet of Things." *McKinsey & Company.* www.mckinsey.com/industries/private-equity-and-principal-investors/our-insights/growing-opportunities-in-the-internet-of-things# (last accessed January 2021).

Dalton, Craig M. (2015) "For Fun and Profit: The Limits and Possibilities of Google-Maps-based Geoweb Applications." *Environment and Planning A* 47(5): 1,029–1,046.

Dalton, Craig M. (2020) "Rhizomatic Data Assemblages: Mapping New Possibilities for Urban Housing Data." *Urban Geography* 41(8): 1,090–1,108.

Dalton, Craig M., and Stallmann, Tim. (2018) "Counter-mapping Data Science." *The Canadian Geographer* 62(1): 93–101.

Dalton, Craig M., and Thatcher, Jim. (2015) "Inflated granularity: Spatial 'Big Data' and Geodemographics." *Big Data & Society* July–December: 1–15.

Dalton, Craig M., and Thatcher, Jim. (2019) "Seeing by Starbucks: The Social Context of Mobile Maps and Users' Geographic Knowledges." *Cartographic Perspectives* 92: 24–42.

Dalton, Craig M., Taylor, Linnet, and Thatcher, Jim. (2016) "Critical Data Studies: A Dialog on Data and Space." *Big Data & Society* January–June: 1–9.

Dalton, Craig M., Wilmott, Clancy, Fraser, Emma, and Thatcher, Jim. (2020) "'Smart' Discourses, the Limits of Representation, and New Regimes of Spatial Data." *Annals of the American Association of Geographers* 110(2): 485–496.

Dance, Gabriel J.X., LaForgia, Michael, and Confessore, Nicholas. (2018) "As Facebook Raised a Privacy Wall, It Carved an Opening for Tech Giants." *The New York Times.* www.nytimes.com/2018/12/18/technology/facebook-privacy.html (last accessed January 2021).

Davies, Harry. (2015) "Ted Cruz Using Firm that Harvested Data on Millions of Unwitting Facebook Users." *The Guardian.* www.theguardian.com/us-news/2015/dec/11/senator-ted-cruz-president-campaign-facebook-user-data (last accessed January 2021).

de Certeau, Michel. (1984) *The Practice of Everyday Life*. Translated by Steven Rendall. Berkeley, CA: University of California Press.

Debord, Guy. (1955) "Introduction to a Critique of Urban Geography." *Les Lèvres Nues* 6. www.cddc.vt.edu/sionline/presitu/geography.html (last accessed January 2021).

Debord, Guy. (1956) "Theory of the Dérive." *Les Lèvres Nues* 9. www.cddc.vt.edu/sionline/si/theory.html (last accessed January 2021).

Debord, Guy. (1967) *Society of the Spectacle*. Translated by Ken Knabb. Wellington, New Zealand: Rebel Press.

Debord, Guy. (1998) *Comments on Society of the Spectacle*. Translated by Malcolm Imrie. London: Verso Books.

Debord, Guy, and Wolman, Gil J. (1956) "A User's Guide to Détournement." *Les Lèvres Nues* 8. https://cddc.vt.edu/sionline/presitu/usersguide.html (last accessed February 2021).

D'Ignazio, Catherine, and Klein, Lauren F. (2020) *Data Feminism*. Cambridge, MA: MIT Press.

Donahue, Phil. (September 6, 1976) *The Phil Donahue Show*. Clip available: www.youtube.com/watch?v=1EwaLys3Zak (last accessed December 2020).

Downey, Gregory J. (2002) *Telegraph Messenger Boys: Labor, Technology, and Geography, 1850–1950*. Abingdon, UK: Routledge.

Durairajan, Ramakrishnan, Barford, Paul, Sommers, Joel, and Willinger, Walter. (2015) "InterTubes: A Study of the US Long-haul Fiber-optic Infrastructure." *ACM SIGCOMM Computer Communication Review* 45: 565–578.

Eades, Gwilym. (2010) "An Apollonian Appreciation of Google Earth." *Geoforum* 41(5): 671–673.

Edwards, J.C. (2007) "The Thinging of the Thing: The Ethic of Conditionality in Heidegger's Later Work." In: Dreyfus, H.L., and Wrathall, M.A. (eds.) *A Companion to Heidegger*. 456–467. Oxford, UK: Blackwell.

Elden, Stuart. (2004) "Between Marx and Heidegger: Politics, Philosophy, and Lefebvre's *The Production of Space*." *Antipode* 36(1): 86–105.

Elegant, Naomi Xu, and McGregor, Grady. (2019) "Hong Kong's Mask Ban Pits Anonymity against the Surveillance State." *Fortune Magazine*. https://fortune.com/2019/10/04/hong-kong-protests-mask-ban-surveillance-anonymity-facial-recognition/ (last accessed January 2021).

Etherington, Darrell. 2020. Strava Raises $110 Million, Touts Growth Rate of 2 Million New Users per Month in 2020. *TechCrunch*. https://techcrunch.com/2020/11/16/strava-raises-110-million-touts-growth-rate-of-2-million-new-users-per-month-in-2020/ (last accessed January 2021).

Eubanks, Virginia. (2018) *Automating Inequality: How High-tech Tools Profile, Police, and Punish the Poor*. New York: St. Martin's Press.

FCC. (2011) "FCC Strengthens Enhanced 911 Location Accuracy Requirements for Wireless Services." *FCC News*. https://web.archive.org/web/201108 11023438/http://transition.fcc.gov/Daily_Releases/Daily_Business/2011/db0712/DOC-308377A1.pdf (last accessed January 2011).

Feenberg, Andrew. (1999) *Questioning Technology*. New York: Routledge.

Feenberg, Andrew. (2002) *Transforming Technology: A Critical Theory Revisited*. Oxford, UK: Oxford University Press.

Foucault, Michel. (1990) *The History of Sexuality: An Introduction*. Translated by Robert Hurley. London: Penguin Press.

Foucault, Michel. (2003) *"Society Must Be Defended": Lectures at the College de France 1975–1976*. London: Picador Press.

Foursquare. (2020) "Monitoring the Effects of COVID-19 with Location Data." *Foursquare Blog*. https://foursquare.com/article/monitoring-the-effects-of-covid-19/ (last accessed December 2020).

Franceschi-Bicchierai, Lorenzo. (2020) "A Tattoo and an Etsy Shirt Led Cops to Arrest Woman Accused of Burning Cop Cars." *Vice*. www.vice.com/en/article/bv8j8w/a-tattoo-and-an-etsy-shirt-led-cops-to-arrest-woman-accused-of-burning-cop-cars (last accessed January 2021).

Friedman, George. (1986) "Eschatology vs Aesthetics: The Marxist Critique of Weberian Rationality." *Sociological Theory* 4(2): 186–193.

Friedman, Monroe. (2017) Using Consumer Boycotts to Stimulate Corporate Policy Changes: Marketplace, Media, and Moral Considerations. In Micheletti, Michele, Follesdal, Andreas, and Stolle, Dietlind (eds.) *Politics, Products and Markets*. 45–62. Abingdon, UK: Routledge.

Fuchs, Christian. (2014) *Digital Labour and Karl Marx*. New York: Routledge.

Gabrys, Jennifer. (2016) *Program Earth: Environmental Sensing Technology and the Making of a Computational Planet*. Minneapolis, MN: University of Minnesota Press.

Galloway, Alexander R., Thacker, Eugene, and Wark, McKenzie. (2014) *Excommunication: Three Inquiries in Media and Mediation*. Chicago, IL: University of Chicago Press.

Gandy Jr., Oscar H. (1993a) *The Panoptic Sort: A Political Economy of Personal Information*. Boulder, CO: Westview Press.

Gandy Jr., Oscar H. (1993b) "Toward a Political Economy of Personal Information." *Critical Studies in Mass Communication* 10: 70–97.

Gane, Nicholas. (2006) "When We Have Never Been Human, What Is to Be Done? Interview with Donna Haraway." *Theory, Culture & Society* 23(7–8): 135–158.

Gaskins, Nettrice R. (2019) "Techno-vernacular Creativity and Innovation across the African Diaspora and Global South." In: Benjamin, Ruha (ed.) *Race, Carceral Technoscience, and Liberatory Imagination in Everyday Life*. 252–274. Durham, NC: Duke University Press.

Georgiadis, Philip, and Beioley, Kate. (2021) "BA Faces Largest-ever Group Privacy Claim in UK over Data Breach." *The Financial Times*. www.ft.com/content/f3dc6c8e-0f65-40d0-a5d5-9d57d3f9d0e0 (last accessed January 2021).

Gladstone, Brooke. (2013) "'If You've Got Nothing to Hide, You've Got Nothing to Fear.'" *WNYC: On The Media*. www.wnycstudios.org/podcasts/otm/segments/260644-if-youve-got-nothing-hide-youve-got-nothing-fear (last accessed January 2021).

Goldberg, Michelle. (2005) "Sexual Revolutionaries." *Salon*. www.salon. com/2005/04/24/satrapi_2/ (last accessed December 2020).

Goldstein, Caroline. (2020) "How One Artist Hacked Google Maps to Fake a Traffic Jam and Make a Point about the Flaws of Big Data." *ArtNetNews*. https://news.artnet.com/art-world/artist-simon-weckert-google-map-hack-1769187 (last accessed January 2021).

Goodin, Dan. (2020) "Beware of Find-My-Phone, Wi-Fi, and Bluetooth, NSA Tells Mobile Users." *Ars Technica*. https://arstechnica.com/tech-policy/2020/08/beware-of-find-my-phone-wi-fi-and-bluetooth-nsa-tells-mobile-users (last accessed January 2021).

Graham, Mark, Hjorth, Isis, and Lehdonvirta, Vili. (2017) "Digital Labor and Development: Impacts of Global Digital Labour Platforms and the Gig Economy on Worker Livelihoods." *Transfer: European Review of Labour and Research* 23(2): 135–162.

Graham, Mark, Hogan, Bernie, Straumann, Ralph K., and Medhat, Ahmed. (2014) "Uneven Geographies of User-generated Information: Patterns of Increasing Informational Poverty." *Annals of the Association of American Geographers* 104(4). https://doi.org/10.1080/00045608.2014.910087 (last accessed July 2021).

Grand View Research. (2018) "Geospatial Analytics Market Size, Share & Trends Analysis Report by Component, by Type, by Application (Surveying, Medicine & Public Safety), by Region and Segment Forecasts, 2019–2025." www.grandviewresearch.com/industry-analysis/geospatial-analytics-market (last accessed January 2021).

Graziani, Terra, and Shi, Mary. (2020) "Data for Justice: Tensions and Lessons from the Anti-Eviction Mapping Project's Work between Academia and Activism." *ACME* 19(1): 397–412.

Greenfield, A. (2006) *Everyware: The Dawning Age of Ubiquitous Computing*. Berkeley, CA: New Riders Publishing.

Gregg, Melissa. (2015) "Inside the Data Spectacle." *Television & New Media* 16(1): 37–51.

Halkort, Monika. (2019) "Decolonizing Data Relations: On the Moral Economy of Data Sharing in Palestinian Refugee Camps." *Canadian Journal of Communication* 44(3): 317–329.

Halpern, Orit. (2014) *Beautiful Data: A History of Vision and Reason since 1945*. Cambridge, MA: MIT Press.

Hao, Karen. (2020) "We Read the Paper that Forced Timnit Gebru out of Google. Here's What It Says." *MIT Technology Review*. www.technologyreview.com/2020/12/04/1013294/google-ai-ethics-research-paper-forced-out-timnit-gebru/ (last accessed December 2020).

Haraway, Donna J. (1988) "Situated Knowledges: The Science Question in Feminism and the Privilege of Partial Perspective." *Feminist Studies* 14(3): 575–599.

Haraway, Donna J. (2016) *Staying with the Trouble: Making Kin in the Chthulucene*. Durham, NC: Duke University Press.

Harman, Graham. (2010) "Technology, Objects and Things in Heidegger." *Cambridge Journal of Economics* 34: 17–25.

Harris, Leila M., and Hazen, Helen D. (2005) "The Power of Maps: (Counter) Mapping for Conservation." *ACME: An International E-Journal for Critical Geographies* 4(1): 99–130.

Harris, Shane. (2012) "Giving in to the Surveillance State." *The New York Times.* www.nytimes.com/2012/08/23/opinion/whos-watching-the-nsa-watchers. html (last accessed July 2021).

Harvey, David. (1999) *Limits to Capital.* New York: Verso Books.

Harvey, David. (2004) "The 'New' Imperialism: Accumulation by Dispossession." *Socialist Register* 40: 63–877.

Harvey, David. (2015) *Seventeen Contradictions and the End of Capitalism.* Oxford, UK: Oxford University Press.

Haslam, Oliver. (2021) "Facebook and Instagram Threaten to Charge for Access on iOS 14.5 Unless You Give Them Your Data." *iMore.* www.imore.com/ facebook-and-instagram-threaten-charge-access-ios-145-unless-you-give-it-your-data (last accessed August 2021).

He, Laura. (2019) "Apple Removes App Used by Hong Kong Protesters to Track Police Movements." *CNN Business.* www.cnn.com/2019/10/10/tech/apple-china-hkmap-app/index.html (last accessed February 2021).

Heidegger, Martin. (1981[1976]) "Only a God Can Save Us." *Spiegel* interview. In: Sheehan, T. (ed.) *Heidegger: The Man and the Thinker.* 45–67. Chicago, IL: Precedent Press.

Heidegger, Martin. (1977a) "Science and Reflection." In: Lovitt, William (trans.) *The Question Concerning Technology and Other Essays.* 155–182. New York: Harper Perennial.

Heidegger, Martin. (1977b) "The Question Concerning Technology." In: Lovitt, William (trans.) *The Question Concerning Technology and Other Essays.* 3–35. New York: Harper Perennial.

Hern, Alex. (2018) "Fitness Tracking App Strava Gives Away Location of Secret US Army Bases." *The Guardian, GPS.* www.theguardian.com/world/2018/ jan/28/fitness-tracking-app-gives-away-location-of-secret-us-army-bases (last accessed December 2020).

Hill, Kashmir. (2012) "The Reaction to 'Girls Around Me' Was Far More Disturbing than the 'Creepy' App Itself." *Forbes.* www.forbes.com/sites/ kashmirhill/2012/04/02/the-reaction-to-girls-around-me-was-far-more-disturbing-than-the-creepy-app-itself/ (last accessed December 2020).

Hill, Kashmir. (2020) "Wrongfully Accused by an Algorithm." *The New York Times.* www.nytimes.com/2020/06/24/technology/facial-recognition-arrest. html (last accessed June 2021).

Hill Staff. (2020) "The Hill's Top Lobbyists 2020." *The Hill.* https://thehill.com/ business-a-lobbying/top-lobbyists/529550-the-hills-top-lobbyists-2020 (last accessed January 2021).

Hillis, Ken, Petit, Michael, and Jarrett, Kylie. (2013) *Google and the Culture of Search.* Abingdon, UK: Routledge.

Hirschorn, Michael. (2010) "Closing the Digital Frontier." *The Atlantic*. www. theatlantic.com/magazine/archive/2010/07/closing-the-digital-frontier/308131/ (last accessed January 2021).

Hogan, Mél, and Vonderau, Asta. (2019) "The Nature of Data Centers." *Culture Machine*. https://culturemachine.net/wp-content/uploads/2019/04/HOGAN-AND-VONDERAU.pdf (last accessed January 2021).

Horkheimer, Max. (1995) *Critical Theory: Selected Essays*. New York: Continuum.

Horkheimer, Max, and Adorno, Theodor W. (2002[1947]) *Dialectic of Enlightenment: Philosophical Fragments*. Translated by Edmund Jephcott. Stanford, CA: Stanford University Press.

Hsu, Jeremy. (2018) "The Strava Heat Map and the End of Secrets." *WIRED*. www.wired.com/story/strava-heat-map-military-bases-fitness-trackers-privacy/ (last accessed December 2020).

Hunt, Sarah. (2014) "Ontologies of Indigeneity: The Politics of Embodying a Concept." *Cultural Geographies* 21(1): 27–32.

Husserl, Edmund. (2001) *Logical Investigations*. Moran, D. (ed.). London: Routledge.

Information Commissioner's Office. (2020) "ICO Fines British Airways £20m for Data Breach Affecting More than 400,000 Customers." https://ico.org.uk/about-the-ico/news-and-events/news-and-blogs/2020/10/ico-fines-british-airways-20m-for-data-breach-affecting-more-than-400-000-customers/ (last accessed January 2021).

Intersoft Consulting. (2016) *General Data Protection Regulation*. https://gdpr-info.eu/ (last accessed January 2021).

Jacobson, Don. (2020) "EU Court Rejects Data-sharing Deal Due to Privacy Concerns." *UPI*. www.upi.com/Top_News/World-News/2020/07/16/EU-court-rejects-data-sharing-deal-due-to-privacy-concerns/7091594894502/ (last accessed November 2019).

Jameson, Fredric. (2003) "Future City." *New Left Review* 21. https://newleftreview.org/issues/ii21/articles/fredric-jameson-future-city (last accessed December 2020).

Jefferson, Brian. (2017) "Digitize and Punish: Computerized Crime Mapping and Racialized Carceral Power in Chicago." *Environment and Planning D: Society and Space* 35(5): 775–796.

Jefferson, Brian. (2020) *Digitize and Punish: Racial Criminalization in the Digital Age*. Minneapolis, MN: University of Minnesota Press.

Jeffries, Stuart. (2016) *Grand Hotel Abyss: The Lives of the Frankfurt School*. London: Verso Books.

Jenkins Jr., Holman W. (2010) "Google and the Search for the Future." *The Wall Street Journal*. www.wsj.com/articles/SB10001424052748704901104575423294099527212 (last accessed January 2021).

Jennings, Michael W. (2008) "The Production, Reproduction, and Reception of the Work of Art." In: Jennings, Michael W., Doherty, Brigid, and Levin, Thomas Y. (eds.) *The Work of Art in the Age of Its Technological Reproducibility and Other Writings on Media*. 9–17. Cambridge, MA: The Belknap Press of Harvard University Press.

Johnston, Craig. (2014) "SAFT: Protecting Our Lines—and Yours—for 30 Years." *Union Pacific Fiber Optic Focus Newsletter* 29(2): 1.

Khatib, Abdelhafid. (1958) "Attempt at a Psychogeographical Description of Les Halles." *Internationale Situationniste* 2. www.cddc.vt.edu/sionline/si/leshalles. html (last accessed January 2021).

Kingsbury, Paul, and Jones III, John Paul. (2009) "Walter Benjamin's Dionysian Adventures on Google Earth." *Geoforum* 40: 502–513.

Kitchin, Rob. (2014) *The Data Revolution: Big Data, Open Data, Data Infrastructures and Their Consequences.* London: SAGE.

Knowledge@Wharton. (2014) "The Surprising Ways that Social Media Can Be Used for Credit Scoring." *Wharton Business Radio.* https://knowledge.wharton. upenn.edu/article/using-social-media-for-credit-scoring/ (last accessed January 2021).

Kobie, Nicole. (2019) "The Complicated Truth about China's Social Credit System." *WIRED.* www.wired.co.uk/article/china-social-credit-system-explained (last accessed January 2021).

Krumm, John. (2011) "Ubiquitous Advertising: The Killer Application for the 21st Century." *Persuasive Computing* January–March: 66–73.

Kuo, Lily. (2019) "China Bans 23m from Buying Travel Tickets as Part of 'Social Credit' System." *The Guardian.* www.theguardian.com/world/2019/mar/01/ china-bans-23m-discredited-citizens-from-buying-travel-tickets-social-credit-system (last accessed July 2021).

Lally, Nick, Kay, Kelly, and Thatcher, Jim. (2019) "Computational Parasites and Hydropower: A Political Ecology of Bitcoin Mining on the Columbia River." *Environment and Planning E.* DOI: 10.1177/2514848619867608.

Lanier, Jaron. (2014) *Who Owns the Future?* New York: Simon & Schuster.

Latour, Bruno. (1993) *We Have Never Been Modern.* Translated by Catherine Porter. Cambridge, MA: Harvard University Press.

Lee, Timothy B. (2017) "Uber Has Only Itself to Blame for the #DeleteUber Campaign." *Vox.* www.vox.com/new-money/2017/2/2/14478044/trump-delete-uber-campaign (last accessed January 2021).

Levenda, Anthony M., and Mahmoudi, Dillon. (2019) "Silicon Forest and Server Farms: The (Urban) Nature of Digital Capitalism in the Pacific Northwest." *Culture Machine.* https://culturemachine.net/wp-content/uploads/2019/04/ LEVENDA-MAHMOUDI.pdf (last accessed January 2021).

Lin, Jialiu, Amini, Shahriyar, Hong, Jason I., Sadeh, Norman, Lindqvist, Janne, and Zhang, Joy. (2012) "Expectation and Purpose: Understanding Users' Mental Models of Mobile App Privacy through Crowdsourcing." In: *UbiComp '12: Proceedings of the 2012 ACM Conference on Ubiquitous Computing* September: 501–510.

Liu, Chuncheng. (2019) "Multiple Social Credit Systems in China." *Economic Sociology: The European Electronic Newsletter* 21(1): 22–32.

Lohr, Steve. (2012) "Sure, Big Data Is Great, but So Is Intuition." *The New York Times.* www.nytimes.com/2012/12/30/technology/big-data-is-great-but-dont-forget-intuition.html (last accessed January 2021).

Loukissas, Yanni Alexander. (2019) *All Data Are Local: Thinking Critically in a Data-driven Society*. Cambridge, MA: MIT Press.

Mackenzie, Adrian. (2017) *Machine Learners: Archaeology of a Data Practice*. Cambridge, MA: MIT Press.

Magid, Larry. (2009) "Facebook's New Policy Makes Users Think about Privacy." *The Mercury News*. www.mercurynews.com/2009/12/10/magid-facebooks-new-policy-makes-users-think-about-privacy/ (last accessed January 2021).

Mann, Steve. (1998) "'Reflectionism' and 'Diffusionism': New Tactics for Deconstructing the Video Surveillance Superhighway." *Leonardo* 31(2): 93–102.

Marcuse, Herbert. (1991[1964]) *One-dimensional Man*. Boston, MA: Beacon Press.

Marcuse, Herbert. (1969) *An Essay on Liberation*. Boston, MA: Beacon Press.

Marcuse, Herbert. (1982) "Some Social Implications of Modern Technology." In: Arato, A., and Gebhardt, E. (eds.) *The Essential Frankfurt School Reader*. 138–162. New York: Continuum.

Marcuse, Herbert. (2009) *Negations: Essays in Critical Theory*. London: MayFly Books.

Martínez, Antonio García. (2019) "No, Data Is Not the New Oil." *WIRED*. www.wired.com/story/no-data-is-not-the-new-oil/ (last accessed January 2021).

Marx, Gary T. (2009) "A Tack in the Shoe and Taking off the Shoe: Neutralization and Counter-neutralization Dynamics." *Surveillance & Society* 6(3), 294–306.

Marx, Karl. (1955[1847]) *The Poverty of Philosophy*. Translated by the Institute of Marxism Leninism. www.marxists.org/archive/marx/works/1847/poverty-philosophy/index.htm (last accessed November 2020).

Marx, Karl. (1990[1848]) *Capital Volume 1*. Translated by Ben Fowkes. London: Penguin Books.

Marx, Karl. (1975[1852]) *The Eighteenth Brumaire of Louis Bonaparte*. Dutt, C.P. (ed.). New York: International Publishers.

Marx, Karl. (1978[1888]) "Theses on Feuerbach." In: Tucker, R.C. (ed.) *The Marx-Engels Reader*. 143–145. New York: W.W. Norton.

Marx, Karl, and Engels, Friedrich. (1978[1848]) "Manifesto of the Communist Party." In: Tucker, R.C. (ed.) *The Marx-Engels Reader*. 469–500. New York: W.W. Norton.

Matsakis, Louise. (2019a) "How the West Got China's Social Credit System Wrong." *WIRED*. www.wired.com/story/china-social-credit-score-system/ (last accessed January 2021).

Matsakis, Louise.(2019b) "The WIRED Guide to Your Personal Data (and Who Is Using It)." *WIRED*. www.wired.com/story/wired-guide-personal-data-collection/ (last accessed January 2021).

Mattern, Shannon. (2017) *Code and Clay, Data and Dirt*. Minneapolis, MN: University of Minnesota Press.

McCarthy, James. (2013) "We Have Never Been 'Post-political.'" *Capitalism Nature Socialism* 24(1): 19–25.

McDonald, Aleecia M., and Cranor, Lorrie Faith. (2008) "The Cost of Reading Privacy Policies." *I/S: A Journal of Law and Policy for the Information Society* 4(3): 543–568.

McKittrick, Katherine. (2021) *Dear Science and Other Stories*. Durham, NC: Duke University Press.

Milan, Stefania, and Treré, Emiliano. (2019) "Big Data from the South(s): Beyond Data Universalism." *Television & New Media* 20(4): 319–335.

Minxi, Zhou. (2019) "The Truths and Myths about China's Social Credit System." *CGTN*. https://news.cgtn.com/news/3d3d774e7751444f32457a6333566d54/index.html (last accessed January 2021).

Mitchell, Andrew J., and Trawny, Peter (eds.). (2017) *Heidegger's Black Notebooks: Responses to Anti-Semitism*. New York: Columbia University Press.

Morrison, Sara. (2020) "How to Make Sure Google Automatically Deletes Your Data on a Regular Basis." *Vox recode*. www.vox.com/recode/2020/6/24/21301713/google-auto-delete-location-youtube (last accessed January 2021).

Mott, Carrie, and Cockayne, Daniel G. (2017) "Citation Matters: Mobilizing the Politics of Citation toward a Practice of 'Conscientious Engagement.'" *Gender, Place & Culture* 24(7): 954–973.

Mozur, Paul. (2018) "Inside China's Dystopian Dreams: A.I., Shame and Lots of Cameras." *The New York Times*. www.nytimes.com/2018/07/08/business/china-surveillance-technology.html (last accessed January 2021).

Nakamura, Lisa. (2014) "Indigenous Circuits: Navajo Women and the Racialization of Early Electronic Manufacture." *American Quarterly* 66(4): 919–941.

Neate, Rupert. (2018) "Over $119bn Wiped off Facebook's Market Cap after Growth Shock." *The Guardian*. www.theguardian.com/technology/2018/jul/26/facebook-market-cap-falls-109bn-dollars-after-growth-shock (last accessed January 2021).

Neff, Gina, and Nafus, Dawn. (2016) *The Quantified Self*. Cambridge, MA: MIT Press.

New York State. (2016) *Senate Bill S6340A*. 2015–2016 Legislative Session. www.nysenate.gov/legislation/bills/2015/s6340/amendment/a (last accessed February 2021).

Nield, David. (2021) "What's Google FLoC? and How Does It Affect Your Privacy?" *WIRED*. www.wired.com/story/google-floc-privacy-ad-tracking-explainer/ (last accessed June 2021).

Noble, Safiya Umoja. (2018) *Algorithms of Oppression*. New York: New York University Press.

Nost, Eric. (2020) "Infrastructuring 'Data-driven' Environmental Governance in Louisiana's Coastal Restoration Plan." *Environment and Planning E*. DOI: 10.1177/2514848620909727.

O'Brien, Sara Ashley. (2020) "The $185 Million Campaign to Keep Uber And Lyft Drivers as Contractors in California." *The Hill*. https://thehill.com/business-a-lobbying/top-lobbyists/529550-the-hills-top-lobbyists-2020 (last accessed January 2021).

O'Connor, James. (1991) "On the Two Contradictions of Capitalism." *Capitalism Nature Socialism* 2(3): 107–109.

O'Connor, James. (1998) *Natural Causes: Essays in Ecological Marxism*. New York: Guilford Press.

O'Neill, Patrick Howell. (2020) "Hackers Build a New Tor Client Designed to Beat the NSA." *DailyDot*. www.dailydot.com/debug/tor-astoria-timing-attack-client/ (last accessed July 2021).

Osler, Jason. (2018) "Flip Phones and Other 'Dumbphones' New Tech Trend." *CBC News*. www.cbc.ca/news/technology/flip-phones-and-other-dumbphones-new-tech-trend-1.4637126 (last accessed January 2021).

Paglen, Trevor. (2016) "Flight Tracking: A Most Unusual Airline." http://vectors.usc.edu/issues/04_issue/trevorpaglen/ (last accessed July 2021).

Pardes, Arielle. (2018) "How to Manage Your Privacy on Fitness Apps." *WIRED*. www.wired.com/story/strava-privacy-settings-how-to/ (last accessed November 2020).

Pasquale, Frank. (2015) *The Black Box Society: The Secret Algorithms that Control Money and Information*. Cambridge, MA: Harvard University Press.

Peluso, Nancy Lee. (1995) "Whose woods Are These? Counter-mapping Forest Territories in Kalimantan, Indonesia." *Antipode* 27(4): 383–406.

Picchi, Aimee. (2018) "Facebook Stock Suffers Largest One-day Drop in History, Shedding $119 Billion." *CBS News*. www.cbsnews.com/news/facebook-stock-price-plummets-largest-stock-market-drop-in-history/ (last accessed January 2021).

Pickles, John. (2004) *A History of Spaces: Cartographic Reason, Mapping, and the Geo-coded World*. New York: Routledge.

Plant, Sadie. (1992) *The Most Radical Gesture: The Situationist International in a Postmodern Age*. London: Routledge.

Posner, Eric, and Weyl, Glen. (2018) *Radical Markets: Uprooting Capitalism and Democracy for a Just Society*. Princeton, NJ: Princeton University Press.

Powell, Victor, and Lehe, Lewis. (2014) "Markov Chains." *Explained Visually*. https://setosa.io/ev/markov-chains/ (last accessed June 2021).

Precarias a la Deriva. (2003a) *A la Deriva, por los Circuitos de la Precariedad Femenina*. Independent film. www.youtube.com/watch?v=WCEsKJrKH9c (last accessed February 2021).

Precarias a la Deriva. (2003b) "First Stutterings of 'Precarias a la Deriva.'" *Caring Labor: An Archive*. https://caringlabor.wordpress.com/2010/12/14/precarias-a-la-deriva-first-stutterings-of-precarias-a-la-deriva/ (last accessed February 2021).

Precarias a la Deriva. (2005) "A Very Careful Strike—Four Hypotheses." Translated by Franco Ingrassia and Nate Holdren. *Caring Labor: An Archive*. https://caringlabor.wordpress.com/2010/08/14/precarias-a-la-deriva-a-very-careful-strike-four-hypotheses/ (last accessed February 2021).

Probrand. (2019) "Cybersecurity Professionals Give up Their Personal Data in Return for … a Doughnut." *Probrand Blog*. www.probrand.co.uk/blog/pb/june-2019/cybersecurity-professionals-give-up-their-personal (last accessed January 2021).

Prudham, Scott. (2009) "Commodification. In: Castree, N., Demeritt D., Liverman, D., et al. (eds.) *A Companion to Environmental Geography*. Oxford, UK: Wiley-Blackwell, pp. 123–142.

Rancière, Jacques. (2009) *The Emancipated Spectator*. Translated by Gregory Elliot. London: Verso Books.

Rheingold, Howard. (1993) *The Virtual Community: Homesteading on the Electronic Frontier*. Reading, MA: Addison-Wesley.

Risen, James, and Lichtblau, Eric. (2005) "Bush Lets U.S. Spy on Callers without Courts." *The New York Times*. www.nytimes.com/2005/12/16/politics/bush-lets-us-spy-on-callers-without-courts.html (last accessed December 2020).

Robb, Drew. (2017) "Building the Global Heatmap." *Medium*. https://medium.com/strava-engineering/the-global-heatmap-now-6x-hotter-23fc01d301de (last accessed December 2020).

Robertson, Morgan. (2004) "The Nature that Capital Can See: Science, State and Market in the Commodification of Ecosystem Services." *Environment and Planning D: Society and Space* 24(3): 367–387.

Rose, Janus. (2016) "FBI Dumps 18 Hours of Spy Plane Footage from Black Lives Matter Protests." *Vice Motherboard*. www.vice.com/en/article/78ke7a/black-lives-matter-spy-plane-fbi (last accessed February 2021).

Russell, Legacy. (2020) *Glitch Manifesto*. London: Verso Books.

Sadler, Simon. (2010) *The Situationist City*. Cambridge, MA: MIT Press.

Sayer, Derek. (1991) *Capitalism and Modernity: An Excursus on Marx and Weber*. New York: Routledge.

Saxenian, Annalee. (1996) *Regional Advantage: Culture and Competition in Silicon Valley and Route 128*. Cambridge, MA: Harvard University Press.

Scahill, Jeremy. (2015) "The Assassination Complex." *The Intercept*. https://theintercept.com/drone-papers/the-assassination-complex/ (last accessed December 2020).

Shen, Lucinda. (2017) "200,000 Users Have Left Uber in the #DeleteUber Protest." *Fortune*. https://fortune.com/2017/02/03/uber-lyft-delete-donald-trump-executive-order/ (last accessed January 2021).

Sheppard, Eric. (1995) "GIS and Society: Towards a Research Agenda." *Cartography and Geographic Information Systems* 22(1): 5–16.

Shewan, Dan. (2017) "Robots Will Destroy Our Jobs—and We're Not Ready for It." *Guardian*. www.theguardian.com/technology/2017/jan/11/robots-jobs-employees-artificial-intelligence (last accessed November 2020).

Siddiqui, Faiz. (2017) "Uber Triggers Protest for Collecting Fares During Taxi Strike against Refugee Ban." *The Washington Post*. www.washingtonpost.com/news/dr-gridlock/wp/2017/01/29/uber-triggers-protest-for-not-supporting-taxi-strike-against-refugee-ban/ (last accessed January 2021).

Silver, Laura. (2019) "Smartphone Ownership Is Growing Rapidly around the World, but Not Always Equally." *Pew Research Center*. www.pewresearch.org/global/2019/02/05/smartphone-ownership-is-growing-rapidly-around-the-world-but-not-always-equally/ (last accessed January 2021).

Situationist International. (1958) "Definitions." In: Debord, Guy (director) *Internationale Situationniste* 1. https://cddc.vt.edu/sionline/si/is1.html (last accessed February 2021).

Smith, Stacey Vanek, and Garcia, Cardiff. (2018) "What It's Like to Be on the Blacklist in China's New Social Credit System." NPR, *All Things Considered.* Available at: www.npr.org/2018/10/31/662696776/what-its-like-to-be-on-the-blacklist-in-chinas-new-social-credit-system (last accessed August 2021).

Stalder, Felix. (2012) "Between Democracy and Spectacle: The Front and Back of the Social Web." In: *Social Media Reader.* New York: New York University Press, pp. 242–256.

Stanley, John, and Hansen, Roz. (2020) "People Love the Idea of 20-minute Neighbourhoods. So Why Isn't It Top of the Agenda?" *The Conversation.* https://theconversation.com/people-love-the-idea-of-20-minute-neighbourhoods-so-why-isnt-it-top-of-the-agenda-131193 (last accessed January 2021).

Statt, Nick. (2021) "WhatsApp Clarifies It's Not Giving All Your Data to Facebook after Surge in Signal and Telegram Users." *The Verge.* www.theverge.com/2021/1/12/22226792/whatsapp-privacy-policy-response-signal-telegram-controversy-clarification (last accessed June 2021).

Steel, Emily, Locke, Callum, Cadman, Emily, and Freese, Ben. (2013) "How Much Is Your Personal Data Worth?" *The Financial Times.* https://ig.ft.com/how-much-is-your-personal-data-worth/ (last accessed January 2021).

Steele, Chandra. (2020) "Know Your (Data's) Worth." *PCMag.* www.pcmag.com/news/know-your-datas-worth (last accessed January 2021).

Stengers, Isabelle. (2018) *Another Science Is Possible: A Manifesto for Slow Science.* Translated by Stephen Muecke. London: Polity Press.

Taneja, Hemant. (2019) "The Era of 'Move Fast and Break Things' Is Over." *Harvard Business Review.* https://hbr.org/2019/01/the-era-of-move-fast-and-break-things-is-over (last accessed November 2019).

Thatcher, Jim. (2013) "Avoiding the Ghetto through Hope and Fear: An Analysis of Immanent Technology Using Ideal Types." *Geojournal* 78: 967–980.

Thatcher, Jim. (2014) "Living on Fumes: Digital Footprints, Data Fumes, and the Limitations of Spatial Big Data." *International Journal of Communication* 8: 1,765–1,783.

Thatcher, Jim. (2016) "The Object of Mobile Spatial Data, the Subject in Mobile Spatial Research." *Big Data & Society.* https://doi.org/10.1177/205395171 6659092 (last accessed January 2021).

Thatcher, Jim. (2018) "Your Data Is You, but It's Not Your Own." *University of Nebraska Press Blog.* https://unpblog.com/2018/02/07/from-the-desk-of-jim-thatcher-your-data-is-you-but-its-not-your-own/ (last accessed November 2019).

Thatcher, Jim, and Dalton, Craig M. (2017) "Data Derives: Confronting Digital Geographic Information as Spectacle." In: Briziarelli, M. and Armano, E. (eds.) *The Spectacle 2.0.* 135–150. London: University of Westminster Press.

Thatcher, Jim, Eckert, Josef, and Shears, Andrew. (2018) *Thinking Big Data in Geography.* Lincoln, NE: University of Nebraska Press.

Thatcher, Jim, O'Sullivan, David, and Mahmoudi, Dillon. (2016) "Data Colonialism through Accumulation by Dispossession: New Metaphors for Daily Data." *Environment and Planning D: Society and Space* 34(6): 990–1,006.

The Economist. (2017) "The World's Most Valuable Resource Is No Longer Oil, but Data." *The Economist.* www.economist.com/leaders/2017/05/06/the-worlds-most-valuable-resource-is-no-longer-oil-but-data (last accessed January 2021).

Thompson, Derek. (2010) "Google's CEO: 'The Laws Are Written by Lobbyists.'" *The Atlantic.* www.theatlantic.com/technology/archive/2010/10/googles-ceo-the-laws-are-written-by-lobbyists/63908/ (last accessed January 2021).

Thompson, E.P. (2008[1978]) *The Poverty of Theory and Other Essays.* London: Monthly Review Press.

Thomson, Iain. (2000) "From the Question Concerning Technology to the Quest for a Democratic Technology: Heidegger, Marcuse, Feenberg." *Inquiry* 43: 203–216.

Thompson, Stuart A., and Wezerek, Gus. (2019) "Freaked Out? 3 Steps to Protect Your Phone." *The New York Times.* www.nytimes.com/interactive/2019/12/19/opinion/location-tracking-privacy-tips.html (last accessed January 2021).

Thoreau, Henry D. (1910) *Walden.* New York: Thomas Y. Crowell.

Thudt, Alice, Hinrichs, Uta, and Carpendale, Sheelagh. (2017) "Data Craft: Integrating Data into Daily Practices and Shared Reflections." In: *CHI 2017 Workshop on Quantified Data & Social Relationships.* Denver, CO.

Treloar, Stephen. (2020) "Norway Halts Coronavirus Tracking App over Privacy Concerns." *Bloomberg.* www.bloomberg.com/news/articles/2020-06-15/norway-halts-coronavirus-tracking-app-over-privacy-concerns (last accessed January 2021).

Tsing, Anna Lowenhaupt. (2015) *The Mushroom at the End of the World.* Princeton, NJ: Princeton University Press.

Tufekci, Zeynep. (2018) "The Latest Data Privacy Debacle." *The New York Times.* www.nytimes.com/2018/01/30/opinion/strava-privacy.html (last accessed November 2019).

Twitter. (2019) *Fiscal Year 2019 Annual Report.* https://s22.q4cdn.com/826641620/files/doc_financials/2019/FiscalYR2019_Twitter_Annual_-Report-(3).pdf (last accessed January 2020).

Twitter. (2020a) "Academic Research: Data on Everything and Anything, at Your Fingertips." https://developer.twitter.com/en/solutions/academic-research (last accessed January 2020).

Twitter. (2020b) "*Investor Fact Sheet.* https://s22.q4cdn.com/826641620/files/doc_financials/2020/q1/Q1_20__InvestorFactSheet.pdf (last accessed August 2021).

Van Grove, Jennifer. (2010) "Twitter Partnership Aims to Inspire Deeper Analysis of Tweets." *Mashable.* https://mashable.com/2010/11/18/twitter-gnip-partnership/ (last accessed January 2021).

Vincent, Nicholas, Hecht, Brent, and Sen, Shilad. (2019) "'Data Strikes': Evaluating the Effectiveness of a New Form of Collective Action against

Technology Companies." *Proceedings of the World Wide Web Conference, May 2019*: 1,931–1,943. https://dl.acm.org/doi/10.1145/3308558.3313742 (last accessed May 2021).

Walljasper, Jay. (2017) "Welcome to the 20-Minute Village." *AARP Livable Communities*. www.aarp.org/livable-communities/livable-in-action/info-2017/20-minute-village/ (last accessed January 2021).

Wang, Yuanyuan. (2013) "China Promotes Beidou Technology on Transport Vehicles." *Xinhua News*. https://web.archive.org/web/20130121184421/http://news.xinhuanet.com/english/sci/2013-01/14/c_132102104.htm (last accessed January 2021).

Wark, McKenzie. (2004) A *Hacker Manifesto*. Cambridge, MA: Harvard University Press.

Wark, McKenzie. (2011) *The Beach beneath the Streets: The Everyday Life and Glorious Times of the Situationist International*. London: Verso Books.

Wark, McKenzie. (2013) *The Spectacle of Disintegration: Situationist Passages out of the 20th Century*. London: Verso Books.

Wark, McKenzie. (2020) *Capital Is Dead*. London: Verso Books.

Weber, Max. (2008[1891]) *Roman Agrarian History and Its Significance for Public and Private Law*. Translated by R. Frank. Claremont, CA: Regina Books.

Weber, Max. (2005[1910]) "Remarks on Technology and Culture." *Theory, Culture & Society* 22(4): 23–38.

Weber, Max. (1946[1919]) "Science as a Vocation." In: Gerth, H.H., and Mills, C.W. (eds.) *From Max Weber: Essays in Sociology*. 129–158. New York: Oxford University Press.

Weber, Max. (2005[1930]) *The Protestant Ethic and the Spirit of Capitalism*. Translated by Talcott Parsons. New York: Routledge Classics.

Weckert, Simon. (2020) "Google Maps Hacks: Performance and Installation, 2020." www.simonweckert.com/googlemapshacks.html (last accessed January 2021).

Wei, Yanhao, Yildirim, Pinar, Van den Bulte, Christophe, and Dellarocas, Chrysanthos. (2016) "Credit Scoring with Social Network Data." *Marketing Science* 35(2): 234–258.

Williams, Raymond. (1973) "Base and Superstructure in Marxist Cultural Theory." *New Left Review* 82: 3–16.

Wilson, Matthew W. (2012) "Location-based Services, Conspicuous Mobility, and the Location-aware Future." *Geoforum* 43(6): 1,266–1,275.

Wilson, Matthew W. (2015a) "Flashing Lights in the Quantified Self-city-nation." *Regional Studies, Regional Science* 2(1): 39–42.

Wilson, Matthew W. (2015b), "Morgan Freeman Is Dead and Other Big Data Stories." *Cultural Geographies* 22(2): 345–349.

Yao, Nathan. (2014) "Where People Run in Major Cities." *FlowingData*. https://flowingdata.com/2014/02/05/where-people-run/ (last accessed December 2020).

Zabou. (2012) *The CCTV Map*. https://thecctvmap.wordpress.com/ (last accessed July 2021).

Žižek, Slavoj. (2009) *The Ticklish Subject: The Absent Centre of Political Ontology.* London: Verso Books.

Zook, Matthew A. (2005) *The Geography of the Internet Industry: Venture Capital, Dot-coms, and Local Knowledge.* Boston, MA: Blackwell.

Zook, Matthew, and Graham, Mark. (2007) "The Creative Reconstruction of the Internet: Google and the Privatization of Cyberspace and DigiPlace." *Geoforum* 38(6) 1,322–1,343. https://doi.org/10.1016/j.geoforum.2007.05.004 (last accessed January 2021).

Index

#DeleteUber 117, 124
20 minute neighborhood 95, 138

abstraction *see* data representations of ourselves
acceptance 5, 7, 71, 72–74
accumulation by dispossession 49, 87
Acxiom 18, 40, 115
AdNauseam 76
Adorno, Theodor W. 134; *see also* Horkheimer, Max
adversarial fashion 83
advertising (targeted) 1, 4, 9, 39, 119–120
AirBnB 111; *see also* Inside AirBnB
algorithm
    algorithmic bias 2, 18, 40–41, 43, 53, 55, 57, 58–59, 120–122, 132
    algorithmic decision making 7, 40, 42, 55, 58–59, 62–64, 70, 120–122, 132
Alibaba Sesame Credit 57
alternatives 2
Amazon 40, 47, 52, 93, 123
    Amazon Sidewalk 131, 139
Amoore, Louise 18, 119, 121–122
Anderson, Chris 28–29, 46–47
Anti-Eviction Mapping Project (AEMP) 112, 124
Apple 43, 52, 73, 81, 85, 86, 131
Automating Inequality 63, 121

BeiDou 82
Beller, Jonathan 51
Benjamin, Ruha 4, 17, 40, 52, 63–64
Benjamin, Walter 4, 13, 15, 18–19, 20–22, 24, 25–26, 32, 53, 88, 95, 107, 134
big data 29, 54, 120, 133; chapter 2

big data divide 48, 83
biopolitics 2.0 39, 56–57
Black Lives Matter 58–59, 109, 116
boycott 113–114
Burrington, Ingrid 78, 136

California Assembly Bill 5 93
California Consumer Privacy Act (CCPA) 92, 115
California Proposition 22 93, 123
Cambridge Analytica 2, 67–68; *see also* Facebook scandals
Capital
capital accumulation 2, 4, 6, 14, 18–20, 30–31, 47–57, 61–62, 86–88, 119–120
    corporeal corkscrew inwards 30, 51, 127
    see also data capitalism, accumulation by dispossession
carceral state 36, 40, 58
CCTV 24, 40, 59, 75, 79
Chaplin, Charlie 13, 15
China
    social credit 57–59, 136
    orientalist black mirror 59–61
    see also HKmap.live
Civic Media 112
Clearview AI 59
collective resistance *see* data contestations
Colonization of everyday life, of lifeworlds *see* data capitalism
contact tracing apps *see* COVID-19
Counter-Cartographies Collective 100–101, 102, 103, 104
counter-mapping 110–113
COVID-19 10, 43, 60–61, 81, 119–120, 126, 136

Daily life *see* everyday life
Data
  Data (combining) 4
  Data (living with) 5, 10, 43, 122
  Data (stand for) *see* data representations of ourselves
  Data abuse narratives *see* data scandals
  Data as commodity *see* data as value
  Data as labor 9, 41, 48, 54–55, 87, 102, 113–116
  Data as value 9, 22, 39–40, 47–50, 54–55, 85, 87–88, 102, 113–115
  Data capitalism 3, 14, 39–42, 47- 57, 61–62, 66, 67–69, 84–85, 86–89, 119–120, 127–128, 131–132
  Data classification 5, 9, 23, 25, 29, 32
  Data colonialism 49–50
  Data contestations 91, *see also* data regulation, data dérive, data détournement, data strike
  Data craft 129
data dérive (drift) 8, 9, 86, 101–105, 124, 138
data detournement 9, 86, 108–113, 124, 127
data, fabrication 75–76
data, faith in 6, 46–47, 52
data, location 1, 49, 91
data, obfuscating 75
data regimes 3, 7, 10–11, 55, 94, 129
data regulation 8, 46–47, 55, 67, 72, 73, 91–94, 111, 115
data, representation of ourselves 1, 4, 12, 22, 34, 39–40, 41–43, 46–47, 50–51, 53–64, 87, 88–89, 119–120, 130
data scandals 1–2, 9–10, 35–38, 40–43, 120–122, 136, *see also* Facebook, scandals, Strava, Home Office
data, speaking for see data representations of ourselves
data, speaking with 46, 50, 61–62, 64, 67–69, 122

Data Feminism 10
data spectacle 7, 36, 54–56, 62, 84–85, 88, 101, 104
data strike 9, 86, 113–118, 124
Debord, Guy 6, 26, 30–31, 33, 53, 88, 90, 95–98, 102, 107–108, 123, 138
dérive (drift) 8, 9, 21, 31, 94–105, *see also* data derive
détournement 9, 105–113, *see also* data detournement
dialectic of hope and fear 2, 5–6, 18–20, 22, 26
Digital OnLine Life and You (DOLLY) 86, 137
Disruption 2, 4, 5, 11, 50, 52
DoorDash 93
Drift see dérive and data derive
drone strikes 2, 22, 42, 135
dystopian accounts of technology *see* dialectic of hope and fear

Electronic Frontier Foundation (EFF) 2, 83
End-user license agreement (EULA) 72–74
enframing 29, 31–32; *see also* Heidegger, Martin
Equifax 115
Escape 6, 8, 66, 71, 81–82, 84
  Escape privilege 66, 81–82, 124
Eubanks, Virginia *see Automating Inequality*
Everyday life 1, 3, 15, 21–22, 51, 53
Everyday resistances *see* tactics; resistance; making present; escape
Experian 58

Facebook 3, 26, 40, 43, 47–48, 66, 76, 84, 85, 92, 93, 103, 115, 123, 131
  Scandals 1, 2, 41–42, 67–69, 74, 116–117
Facial recognition 1, 2, 59, 75
Fear *see* dialectic of hope and fear

Federal Communications
    Commission (FCC) 81–82
Film 16–18, 135
Foucault, Michel 39, 56, 65
Foursquare 41–42, 43, 135–136
Frankfurt School 5, 15, 23–25, 31–33,
    133, 134
Franklin, Benjamin 34
Friedman, Milton 24

Gabrys, Jennifer 39–40, 56
Gandy, Oscar Jr. 40–41
García Martínez, Antonio 47
Gaskins, Nettrice R. 25
General Data Protection Regulation
    (GDPR) 8, 91–92, 93, 94, 115
Gebru, Timnit 28
Genomic profiling 2
Geodemographics see data classifi-
    cation
Gig-economy 18, 6, 93, 98, 111, 115,
    123, 128, see also precarious
    labor
Girls Around Me 41–42
Github 12, 44, 109, 128, 138
Glitch 1, see also technology, roles of
    users
Global Positioning System (GPS) 1, 2,
    27, 35, 36, 41, 53, 58, 66, 67, 70,
    75, 76, 82, 104, 108, 135
GNIP see Twitter
God 29–30, 31
Google 2, 28, 69, 85, 92, 93, 103, 104,
    115, 123
    Artificial intelligence 28, 52
    Maps 12, 17, 43, 51, 65–66
Gregg, Melissa 36, 54–56; see also data
    spectacle
Grindr 42, 74

Hacking see resistance
Halpern, Orit 39
Haraway, Donna 3, 4, 13, 18, 23, 108
Heatmap 1; see also Strava
Heidegger, Martin 26–30, 31, 32, 33,
    123, 134

HKmap.live 113, 138
Home Office (U.K.) 10, 37–38, 135
Homelessness 10
Hope see dialectic of hope and fear
Horkheimer, Max 23–25, 33, 34

IBM 18
Individual that capital can see see data
    representations of ourselves
Infrastructure 50–51, 78
Inside AirBnB 111, 124
Instacart 127
Instagram 24, 62, 131
Institute of Applied Autonomy 79,
    137
Internet of things (IoT) 33

Jefferson, Brian 17, 18, 36, 133

Kalanick, Travis see #DeleteUber
kettling 125, 127, 138, see also Sukey
Khatib, Abdelhafid 94–95, 96–98
Kimura, Hana 137

La dialectique peut-elle casser des
    briques? 105–106
LinkedIn 16
Lived cultures of technology see
    everyday life
Living in the cracks 9, 126–128
Living with data see data, living with
Lobbying 93
Lyft 93, 117

Making present 6, 8, 71, 77–80
Malcolm, Ian 119
Mapping Police Violence 112
Marcuse, Herbert 14, 26, 31–33, 122,
    123
Margin of maneuver see technology
    roles of users
Marx, Karl 6, 14, 18–20, 23, 24, 109,
    133, 134
Marxism 6, 14, 18–20, 30, 31
    Superstructure 23–25
Media cycle see data scandal

Microsoft 1, 47, 93, 114

National Security Agency (NSA) 42–43
Noble, Safiya 63

Occupy Wall Street 116, 125
Ofqual 120–121
Oil, as metaphor for data 46–49
One-dimensional man *see* Marcuse, Herbert

Palantir 115
Parler 113
Partnership on AI 2
Pokémon GO 115
Policy 2–3, 93
Post-political 125–126
Praxis 3, 9, 102–103, 110–113, 122, *see also* resistance (active), making present, escape, dérive (drift), détournement, data strikes
Precarias a la Deriva 8, 9, 98–100, 101, 102, 104, 105, 114, 116, 118, 124
Precarious labor 98–100
Prediction 4, 52, 132
Press coverage *see* data scandal
Privacy *see* surveillance vs. privacy
Privacy-washing 66, 67–69
Profiling *see* data classification
Profit-seeking *see* capital accumulation, data capitalism
Psychogeography *see* dérive (drift)

Quantification 5, 14, 23–24, 32, 34, 39–40, 61

Ranking *see* data classification
Reeves, Keanu 81
Regulation *see* data regulation
Resistance 3, 4, 5, 65–66
  Resistance (active) 5, 7–8, 71, 74–77, 84

Schmidt, Eric 52, 69, 136

Security vs. privacy *see* surveillance vs. privacy
Signal (app) 116
Silicon Valley 5, 6, 7
Situationist Internationale 8,9, 30–31, 53, 83, 90–91, 94–98, 99, 104, 105–108, 116; *see also* Debord, Guy
Snapchat 12, 43
Snowden, Edward 42, 43
Social credit *see* China, social credit
Solidarity 99–101
Sousveillance 108
Spatial data regimes *see* data regimes
Spectacle 30–31, 53, 88–89; *see also* data spectacle
Springsteen, Bruce 23, 134
Strava 1, 35–37, 38, 43
Strike 9; *see also* data strike
Student Nonviolent Coordinating Committee 116
Sukey 113
superstructure 137; *see also* Marxism, superstructure
surge pricing 117
Surveillance
  By police 58–61
  Vs. privacy 66, 69–71

Tactics 65–67, 71, 75–76, 82–83, 85
Technological determinism 15, 19–20, 26, 133–134
Technological neutrality 15–16, 27
Technology
  And capitalism 13–14, 30–33
  Hope and fear of *see* dialectic of hope and fear
  Roles of designers 13, 16–18, 69
  Roles of users 16–18, 43, 69–71
  As standing-reserve 28–29, 32, 39; *see also* Heidegger, Martin
Telegram 116
Telephone 13, 18, 20–21, 84
Temperature blankets 128–129
TikTok 12, 93
Tinder 40, 41–42, 74, 76

Tor Project 76–77
TransUnion 58
Tsing, Anna Lowenhaupt 1, 3
Twitter 4, 74, 86–88, 109, 115, 123, 137
Typology of responses to data
    capitalism 71–83; see also
    acceptance, resistance (active),
    making present, and escape

Uber 2, 18, 40, 81, 83, 93, 117, 123; *see*
    also #DeleteUber
Union Pacific 50
Unities of ambiance see dérive (drift)

Utopianism and technology *see*
    dialectic of hope and fear

Wark, McKenzie 3, 90
Weber, Max 19–20, 23, 24, 134
Weckert, Simon 65
WhatsApp 116–117
Wilson, Matt 34

Yelp 27, 127

Zuckerberg, Mark 2, 26, 68; *see also*
    Facebook, scandals

Thanks to our Patreon Subscribers:

Lia Lilith de Oliveira
Andrew Perry

Who have shown generosity and
comradeship in support of our publishing.

Check out the other perks you get by subscribing
to our Patreon – visit patreon.com/plutopress.

Subscriptions start from £3 a month.

Made in the USA
Coppell, TX
17 January 2023

11239637R00109